Nick Paul

SCIENCE for You

Biology

Series editor
Lawrie Ryan

Published in 2002 by:
Nelson Thornes Ltd
Delta Place
27 Bath Road
CHELTENHAM
GL53 7TH
United Kingdom

05   06   07   08   /   10   9   8   7   6   5   4   3   2

A catalogue record for this book is available from the British Library

ISBN 0 7487 6693 6

Illustrations and page make-up by Wearset Ltd, Boldon, Tyne and Wear

Printed and bound in China by Midas Printing International Ltd

Introduction

Science for You (Biology) has been designed to help students studying Double or Single Science at Foundation level.

The layout of the book is easy to follow, with each new idea on a fresh double page spread. Care has been taken to present information in an interesting way. You will also find plenty of cartoons to help you enjoy your work.
Each new biological word is printed in bold type and important points are in yellow boxes.

There are short questions in the text, as well as a few questions at the end of each spread. These help you to check that you understand the work as you go along. The questions at the end of the chapter are there to encourage you to look back through the chapter and apply your new ideas. At the end of each section, you will find lots of past paper questions to help with revision. These are on the coloured pages throughout the book.

At the end of each chapter, you will see a useful summary of the key facts you need to know. You can test yourself by answering Question 1 that follows each summary.

As you read through the book, you will come across these signs:

This shows where there is a chance to use computers to help you find information or view simulations.

This shows where experiments can be done to support your work.

(The instructions are on sheets in the Teacher Support CD ROM.)

There is an extra section at the end of the book.
Here you can get help with your Coursework, Revising and doing your exams, and Key Skills.

Using this book should make biology easier to understand and bring you success in your exam.

Finally, I hope you'll have fun studying biology, after all, most of us enjoy the things we're good at!

Good luck!

Nick Paul

Contents

CHAPTER 1

Cells

▶▶▶ 1a Building blocks

Look out of your window. Most of the buildings
you can see are probably made of building
blocks called bricks.

All living things are made of building blocks called **cells**.

You need a microscope to see cells. But just because
they are so small don't think that they are not important.

Look at the amazing photograph below:
It shows a small bundle of cells under a powerful
microscope.
This is a **human embryo**. It is the first stage
in the growth of a new human being.

You have probably heard of test tube babies.
This is where scientists take a female egg cell and
mix it with male sperm cells in a small dish.
If one of the sperms joins with the egg then an embryo
will grow.
To do this scientists must know a lot about how cells work.

Putting the blocks together.

a) What do we call it when the sperm and egg cells
join together?

What else can we do with cells?

Scientists can now grow human skin cells to help with
skin grafts. These are needed for people who have had
a lot of their skin burned away in a fire.

They can also use cells to grow complete plants.
If kept in the right conditions a small group of cells taken from a plant
can grow into a new plant. This new plant will have
all the good features of the original plant.

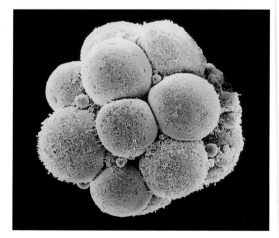

A human embryo.

How do we know that cells exist?

Cells can only be seen with a microscope.
One of the first microscopes was produced
by a British scientist called Robert Hooke.
In 1665 Hooke used his microscope
to look at strips of cork.

b) Where does cork come from?

An early microscope.

He saw lots of tiny box-like shapes
which he called cells.
It took nearly two hundred years for scientists
to appreciate how important Hooke's work was.
In 1838 two German scientists suggested
that all living things were made of cells.

Today the microscopes you use in school
probably look like this one:

c) Why do you think this is called a light microscope?

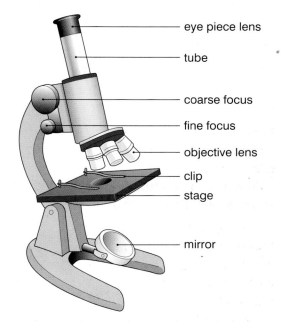

- eye piece lens
- tube
- coarse focus
- fine focus
- objective lens
- clip
- stage
- mirror

Most school microscopes can magnify
up to about 400×.
This makes an object look 400 times bigger
than it really is.
This is more than enough to see the basic
structure of a cell.
There is a much more powerful microscope.
The electron microscope can magnify up to
500,000 times!
This uses beams of electrons instead of light.
The image appears on a TV screen, instead of
through an eyepiece.
Electron microscopes are very expensive
but they do show far more detail.

Remind yourself!

1 Copy and complete:

All living things are made of ......

These can only be seen with a ......

New ...... can be grown to provide ...... grafts for
burn victims. In the right conditions whole ......
can be grown from just a few cells.

2 Explain why scientists had not seen cells before
1665.

3 Find out about Robert Hooke. How did he come
up with the name 'cells'?

On page 6 we saw that cells are
'the building blocks of life'.

But obviously they are a bit more complicated than bricks.
A typical animal cell has these features:
- a **nucleus** – this controls everything that happens
  inside the cell, and contains inheritance information.
- **cytoplasm** – this is a jelly like substance. Lots
  of chemical reactions take place here.
- a **cell membrane** – this is the thin layer on the outside.
  It controls what enters and leaves the cell.

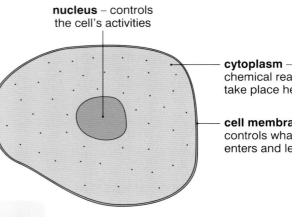

nucleus – controls
the cell's activities

cytoplasm –
chemical reactions
take place here

cell membrane –
controls what
enters and leaves

a) Where do you usually find the nucleus in an animal cell?

A typical plant cell looks a bit different.
This is because as well as a nucleus, cytoplasm and cell membrane,
it also has:
- a **cell wall** – this gives the cell strength.
- **chloroplasts** – these are special structures in the cytoplasm.
  Their job is to trap sunlight to help the plant make food.
- a **vacuole** – this is a large sac filled with a watery
  liquid called cell sap. When the sac is full it helps
  the cell to stay firm.

b) What important process happens in chloroplasts?

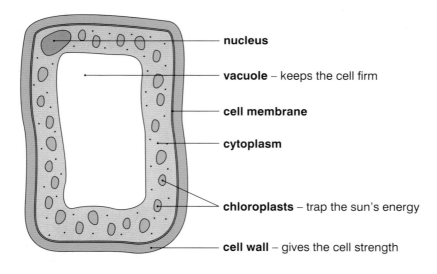

nucleus

vacuole – keeps the cell firm

cell membrane

cytoplasm

chloroplasts – trap the sun's energy

cell wall – gives the cell strength

What do cells really look like?

You've probably guessed by now that *typical* cells
don't really exist.
The drawings on the last page are a way
of clearly showing you all the parts.

Scientists have used microscopes like those on page 7
to produce these drawings.

Here are some photographs of cells under a microscope:

This one is from some human cheek cells.

a) What do you notice about them?
Are they all the same shape?

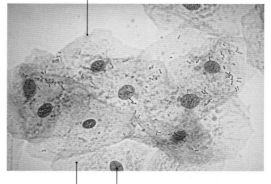

cell membrane

cytoplasm   nucleus

Unlike the cells inside your cheek these cells look blue.
This is because they have been stained.

b) Why do you think the cells were stained?

This photo shows cells from an onion.

Look carefully at the cells:

These cells have a more regular shape than the cheek cells.

Again these cells have been stained to make them easier to
see.

You will be able to look at cells like
these using a microscope.
When you do, try to draw them but
remember these simple rules:
- use a pencil
- only draw two or three of each type
- label the different parts
- draw what you see under the microscope
  not what you see in the photograph.

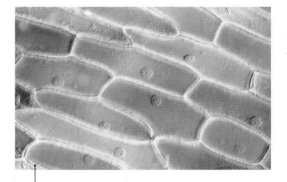

cell wall

Remind yourself!

1 Copy and complete:

   We need a ...... to see cells clearly.

   A cell contains a ...... which controls its activities.
   The ...... is where all the chemical reactions take
   place. The cell membrane controls what ......
   and ...... the cell.

2 Write down three differences between the plant
   and animal cells in the photographs.

3 It's a fact that our skin cells are being rubbed off
   all the time.
   So why don't we just wear away?

▶▶▶ 1c Different cells for different jobs

We have already seen that animal and plant cells
are different from each other.
But there are, in fact, lots of different animal cells.

a) Why do we have different types of cell?
What different jobs do they do?

Imagine that you are a nerve cell connecting your
big toe to your brain.
What shape would you need to be?
Long and thin might be a good idea, shaped like a wire.

Now let's think about a sperm cell?
This is a very small cell that has to swim a very long
way.

Look at its special features below:
● a long tail to help it swim
● a streamlined shape
● a middle section that makes lots of energy.

b) Why do you think sperm cells need a lot of energy?

As you work through this book you will meet
other animal cells that are **adapted** to do a certain job.

For example:
● red blood cells with no nucleus that can
pack in loads of oxygen
● white blood cells that can surround and destroy
bacteria
● muscle cells that can contract (shorten).

Animals that are made up of lots of
different cells are called **multicellular**.

The simplest living organisms that are
made of just one cell are called **unicellular**.

You can read about a unicellular organism
on page 12.

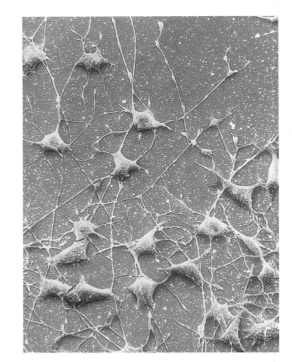

Nerve cells are long and thin.

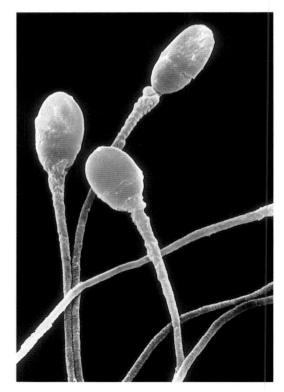

Sperm cells have long tails.

Do plants have lots of different cells too?

The answer is yes and for exactly the same reason.
Different plant cells do different jobs.

Let's look at two examples:

These are **leaf palisade** cells.
They are found near the top which is the bit of the
leaf that receives most light.
They are packed full of chloroplasts to
trap all of this useful sunlight.

Another example is the **root hair** cell.
If you ever look at young plants (like cress seedlings)
you will notice tiny hairs on their roots.
These hairs are, in fact, cells.
Their job is to make the surface of the root even bigger
than normal.

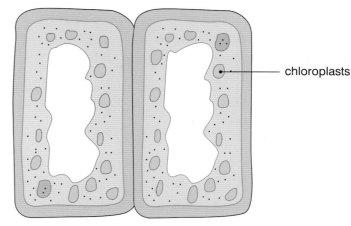

chloroplasts

leaf palisade cells are
packed full of chloroplasts

c) Why is this useful?

d) Why must you be careful when transferring seedlings from seed
trays into separate pots?

Young plants need to take in a lot of water to help them grow.
It is through the roots that this water enters the plant.

There are lots of other examples.
Later in this book you will:

- Look at hollow cells that join up to make long tubes.
  These help plants to draw water up to the leaves.
  These cells are called **xylem** cells.

- Another group of cells that form tubes are called **phloem cells**.
  These tubes transport food to all parts of the plant.

- You will also find cells in the leaves that open and close tiny
  holes. These cells allow gases and water vapour to pass through.
  The holes are called **stomata**, and the sausage shaped cells that
  open and close them are called **guard cells**.

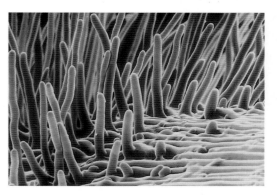

Tiny root hairs on a seedling.

Remind yourself!

1 Why are cells not the same?

2 Why do sperm cells have tails?

3 How are nerve cells different from cheek cells?

4 Why do leaf palisade cells contain lots of
chloroplasts?

5 Why do young plants need roots with a big
surface area?

Humans contain billions of cells.
Obviously all these cells do not work on their own.

> Groups of identical cells that all do the same job
> are called **tissues**.

For example heart muscle cells make special heart tissue that beats throughout your life.

Brain tissue contains millions of nerve cells.
These sort out and store information.

> A number of different tissues join together to make
> an **organ**.

For example the heart is made of muscle tissue, blood tissue and nerve tissue.

All these tissues do different jobs but they work as a team as the heart pumps blood around the body.

The stomach is another important organ.
It helps to digest the food you eat.

> A number of organs working together make up a **system**.

For example the heart and blood vessels make up the circulatory system.

Put all the systems together and what do you get?
You! of course.

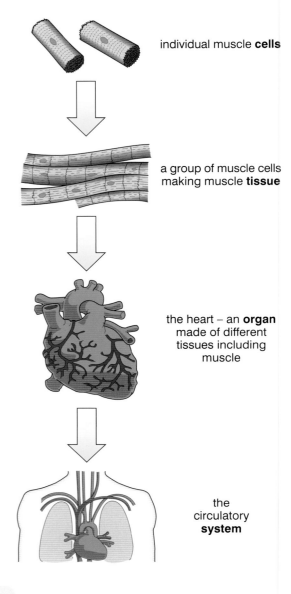

individual muscle **cells**

a group of muscle cells making muscle **tissue**

the heart – an **organ** made of different tissues including muscle

the circulatory **system**

> **a)** Now make a list of all the systems you have heard of in the human body.

Some living things are far simpler.
Creatures like the **amoeba** are made of just one single cell.
This creature does not have tissues, organs and systems, but it is still capable of normal life.
For example it can feed, exchange gases, move and also reproduce itself.

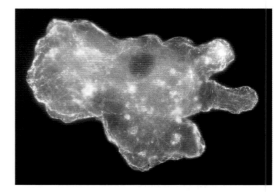

An amoeba.

In/out, in/out

Why do smoke alarms go off even
if the fire is in another room?

The smoke is spreading by a process
called **diffusion**.

> **Diffusion is the movement of particles from a high concentration to a
> low concentration until they are spread out evenly.**

Diffusion is how useful substances get into
cells. It is also how they get rid of some
waste materials.

Smoke diffuses up to the alarm.

> **b)** What do you think cells need to take inside?

Food and oxygen are needed to make energy.
They are carried by the bloodstream to every cell
in your body.

They get into the cell by **diffusing** through the **cell
membrane**.

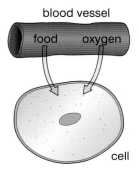

> **c)** Which part of the cell do you think they enter?
> (Hint: look back at page 7.)

Cells make waste materials like carbon dioxide.
These materials must be removed before they harm
the cells.
Carbon dioxide and other waste materials diffuse
out of the cell.
They move *from* the cell *into* the blood.

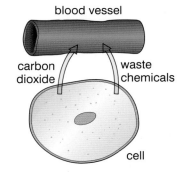

Remind yourself!

1 Copy and complete:

The heart is an example of an ......

An organ is a collection of ...... working together.
A ...... is made up of a number of organs
working together.

2 What things diffuse:
   a) into a cell?
   b) out of a cell?

3 Which bit of an animal cell do these materials
have to diffuse through?

Summary

All living things are made of **cells**.
Most cells have a **nucleus**, **cytoplasm** and **cell membrane**.
Plant cells also have a **cell wall**, **vacuole** and **chloroplasts**.

Different cells are designed to do different jobs.

Tissues are groups of cells that all do the same job.
An **organ** is made up of a number of tissues working together.
Different organs join together to form a **system**.

Useful substances enter cells by **diffusion**.

Questions

1 Copy and complete:

All living things are made of ....

These are so small we need a ...... to see them. Most cells contain a ...... which controls what happens ...... the cell. They also contain ...... and a .... Plant cells also contain ...... which trap the Sun's energy. The plant has a .... which gives the cell strength.

2 Here is a drawing of a human cheek cell.

a) Label parts A, B and C.

b) What is the job of parts B and C?

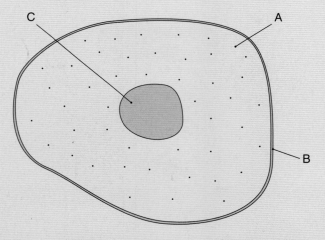

3 Draw a mind map to show all you know about cells.

4 This is a cell from a plant:

a) Label parts A, B and C.

b) Which two parts are only found in plant cells?

c) What is the job of part C?

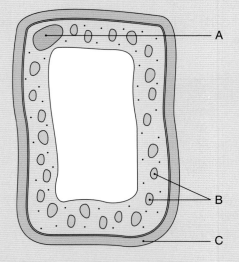

5 Look back at the drawings on page 8.

Now copy and complete this table to show three differences between plant and animal cells.

	Plant	Animal
1		
2		
3		

Section One
Humans as organisms

In this section you will find out about three important systems in the human body, digestion, breathing and circulation. You will also find out about how the body defends itself against infection.

CHAPTER 2

FOOD

▶▶▶ **2a Food**

What do people mean when they talk about a diet?
Usually it means going without food so you can lose
weight. But to doctors and scientists it means
something else. The **human diet** is everything
we eat.

a) Write down your diet for a typical school day.

For healthy living, it is important that our diet is **balanced**.
A balanced diet must include a number of different foods.
In particular we must eat the right amounts of:
- **Carbohydrates** (starch and sugar)
- **Proteins**
- **Fats**.

Also we need small amounts of **minerals** and **vitamins**,
together with **roughage (fibre)** and plenty of **water**.

b) Can you remember why we need these foods,
and where we can get them from?
Do some research and complete the table below.

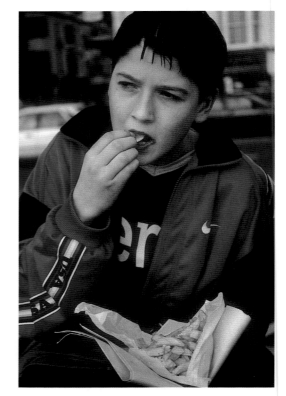

'Is this your daily diet?'

Food type	Why is it needed?	Which food is it in?
Carbohydrates		
Proteins		
Fats		
Minerals		
Vitamins		
Roughage		

What about water?
Water is important because all the chemical
reactions in your body take place in solution.

Water in fact makes up something like 70% of your body!

c) Now look back at your answer to question a).
Does your daily diet look balanced?

How much is enough?

It's all very well being told to eat a balanced diet, but just how much food do we need?

The amount that a person needs depends on a number of factors, such as:

- age
- whether the person is growing
- how active they are.

d) Try to list some other factors.

All of these factors influence how much energy you need. But energy is not the only reason we need food.

e) Try to list some other reasons.

Energy is usually measured in units called kilojoules (kJ). This table compares the daily energy needs of a number of different people.

| | Daily energy needs (kJ) | |
	Male	Female
8-yr old	8500	8500
Yr 10 pupil	12 500	9800
Science teacher	11 000	9700
Adult manual worker	15 000	12 500

f) Why do you think the pupil needs more energy than the teacher? (Two possible reasons)

g) Why do boys and girls of the same age need different amounts of energy?

Today you can buy drinks especially designed to give you an energy boost.

Remind yourself!

1 Copy and complete:

A balanced diet must include fats ...... and ...... Minerals and ...... are also important. ...... makes up 70% of our body.

Food gives us ...... The amount of energy we need depends on our ...... and how ...... we are.

2 Draw a mind map to show all you know about the human diet.

3 Find out about the deficiency disease called scurvy.

Do you ever eat until you're absolutely full? A full stomach doesn't mean that the food inside is doing you any good. To benefit from the goodness in food it has to be in your bloodstream. It can then be carried to all your cells.

Obviously a whole chip or a chunk of chocolate will not fit into your blood vessels.

First of all it has to be **digested**.

> **Digestion is the breaking down of food into small molecules that dissolve easily.**

When a meal has been digested it can then enter the bloodstream. This is a process called **absorption**.

All this happens in part of your body called the **digestive system** (gut).

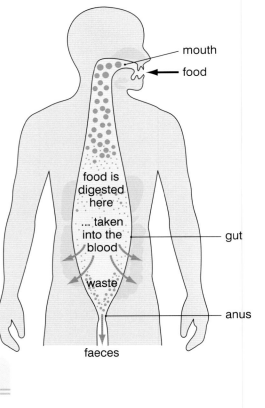

a) Write down a list of all the parts of this system that you can remember.

The digestive system is basically a 6 m-long tube. It starts with your mouth and ends at your anus.

b) How tall are you (in metres)?

No doubt you are a lot less than 6 m tall.

c) Look at the drawing of the digestive system on the next page. Explain how it can fit into your body.

Food can take up to three days to pass right through you. It depends on the type of meal you have eaten. The digestive system has to be so long because some meals take so long to digest.

The digestive system

On this page we will look at a drawing of
the human digestive system.

Don't worry, you will not have to draw this in an exam!
But as we learn more about digestion it will help you
to know where the main parts are and what they do.

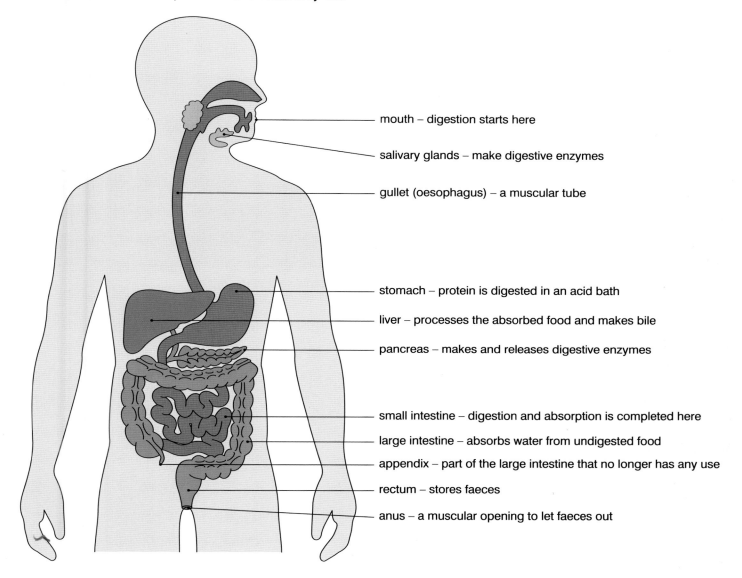

mouth – digestion starts here

salivary glands – make digestive enzymes

gullet (oesophagus) – a muscular tube

stomach – protein is digested in an acid bath

liver – processes the absorbed food and makes bile

pancreas – makes and releases digestive enzymes

small intestine – digestion and absorption is completed here

large intestine – absorbs water from undigested food

appendix – part of the large intestine that no longer has any use

rectum – stores faeces

anus – a muscular opening to let faeces out

Remind yourself!

1 What does the word digestion mean?

2 Why does food need to be digested?

3 When food is absorbed what happens to it?

4 Why is the digestive system so long?

5 Where does digestion begin?

6 How does food move along the gut?

7 What is undigested food called?

How does food get broken down?
The process starts in the **mouth**.
The teeth chop up food into small pieces.
However, the average chip isn't in your
mouth for more than a second or two.

- Most digestion is carried out by
  chemicals called **enzymes**.
- Enzymes break **large insoluble**
  chemicals into **small soluble** ones.

Digestive enzymes are made by cells in the
wall of the gut. Their job is to speed up
the digestion of food.

> **a)** In chemistry what do we call chemicals
> that speed up reactions?

Using enzymes

Enzymes are **not** living but they are very
important to living things.
They have also become very useful to us in
lots of other ways.

Have you ever thought about why some washing
powders are called 'Bio'?
It is because they contain enzymes to help
break down food stains.

Enzymes are also used in cheese making.
Cheese is made from the solid part of clotted milk.
Enzymes called proteases are used to clot the milk.

Enzymes are also responsible for those
chocolates that are gooey inside.
The inside starts off as a solid but is then
broken down by enzymes.

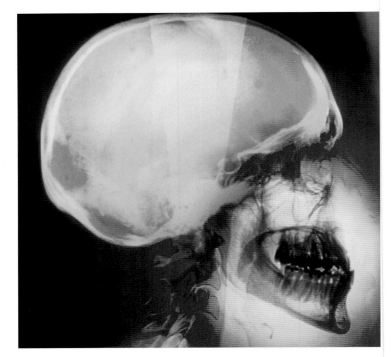

Digestion starts here in the mouth.

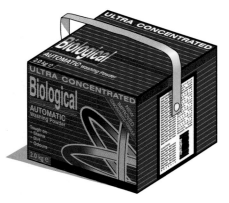

A common use of enzymes.

The right enzyme for the job

In your body there is one type of enzyme for each of the three main food types.

b) What are the three main food types in the human diet?

Carbohydrases (e.g. amylase) are made in the **salivary glands, pancreas** and **small intestine**.
They break down **starch** into **simple sugars**.

a starch molecule is made up of lots of glucose molecules

glucose

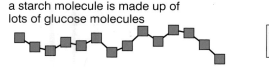

carbohydrase enzyme

Proteases are made in the **stomach, pancreas** and **small intestine**.
They break down **proteins** into **amino acids**.

a protein molecule is made up of lots of different amino acids

amino acids

protease enzyme

Lipases are made in the **pancreas** and **small intestine**.
They break down **fats** into **fatty acids** and **glycerol**.

a fat molecule is made of fatty acid and glycerol

glycerol — fatty acids

lipase enzyme

glycerol   fatty acids

Remind yourself!

1 How do enzymes help digestion?

2 Where are enzymes made?

3 In which part of the gut are all types of enzyme found?

4 Put the information on this page into a summary table like this:

Large molecule	Enzyme	Small molecule

▶▶▶ 2d The right conditions

Enzymes are a bit like some people – awkward!

To be precise, each type of enzyme works best in just the right conditions. This usually means either acidic or alkaline.

A good example of this is in the stomach.

> **a)** Which type of enzyme is found in the stomach?
> (Look at the diagram on page 21.)

The stomach contains very strong acid. You will know this because when you are sick your throat stings. The protease enzymes found here work very well in these conditions. But put them in the alkaline small intestine and they will not work at all.

> **b)** What is the other use of the strong acid in the stomach?
> (Hint: it has something to do with protection from illness, look in Chapter 5.)

Think about this:
If the gut is one long tube, how does strong acid in the stomach, suddenly become alkaline in the small intestine?

This is where **bile** comes in.
Bile is made in the **liver** and stored in the **gall bladder**.

It enters near the top of the small intestine.
Here it does two important jobs:

1 it **neutralises** the strong stomach acid.

This is necessary because unlike the stomach the enzymes here work best in alkaline conditions, and

2 it helps to break up large fat globules into smaller ones (a process called **emulsification**).

Bile is **not** an enzyme but it makes the job of lipase enzymes easier.

If the large fat globules are broken up into smaller ones there will be a bigger surface area of fat.
This means that the fat will be digested more quickly.

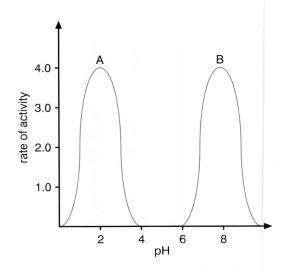

Which enzyme works best in acid conditions A or B?

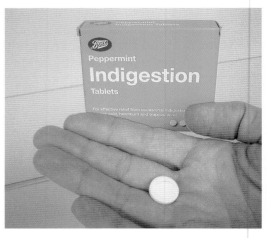

Indigestion tablets do a similar job to bile.

Absorption

The small molecules of food are of no use
until they are absorbed into our bloodstream.

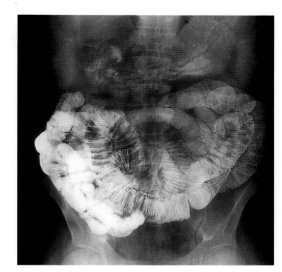

c) What are the three small food molecules
called? (Hint: look on page 21.)

Absorption takes place in the small intestine.
Here the molecules pass through its thin wall.
This happens when these molecules move from a
high concentration in the gut, to a lower concentration
in the blood.

d) What do we call this type of movement?
(Hint: look back to page 13.)

Food is absorbed here.

The small intestine is well designed for absorbing food.
It has these special features:
- it is very long so it has a **big surface area**
- it has a **thin wall** allowing food through easily
- it has a **good blood supply** to receive the food
- it has thousands of **villi** which further increase
  its surface area.

What are villi?

Villi are tiny finger-like projections that stick out
into the middle of the gut. Each villus can absorb food
particles.

And finally!

The food that cannot be digested passes into the large intestine.
Here most of the **water** left in it is absorbed as this is an
important part of the diet.
What remains is called **faeces**, and that is flushed down the toilet.

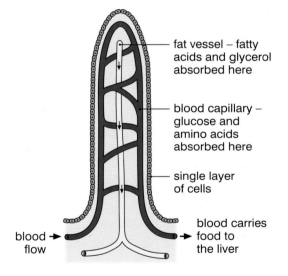

fat vessel – fatty
acids and glycerol
absorbed here

blood capillary –
glucose and
amino acids
absorbed here

single layer
of cells

blood
flow

blood carries
food to
the liver

Remind yourself!

1 Copy and complete:

Inside the stomach there is strong …… This
helps the enzymes and also kills ……. Bile ……
the stomach acid and helps the …… enzymes.
Villi give the small intestine a …… surface area.
This makes …… easier. Undigested food is
called ……

2 Why do we need bile to break up fat particles?

3 How do villi give the small intestine such a big
surface area?

Summary

The three main food types are carbohydrates, proteins and fats.

Digestion involves breaking down food into small soluble molecules.
Digestion is speeded up by special chemicals called **enzymes**.
These chemicals are made in the salivary glands, the pancreas and the small intestine.

Carbohydrases break down **carbohydrates** into **sugars**.
Proteases break down **proteins** into **amino acids**.
Lipases break down **fats** into **fatty acids** and glycerol.

Different enzymes need different conditions to work in.
The inside of the stomach is acidic. The acid helps the enzymes to break down protein. It also kills bacteria.

Bile is made in the liver and it is stored in the gall bladder.
The inside of the small intestine is alkaline. Bile neutralises the stomach acid and helps the lipase enzymes to digest fat.

Digested food passes from the small intestine into the bloodstream.
This process is called **absorption**.

Undigested food passes into the large intestine. Most of the water is absorbed from it. It is now called **faeces**.

Questions

1 Copy and complete:

Starch and sugar are examples of ......
Proteins are needed for ...... and fats give us ......

Digestion breaks down food into ...... particles.
These are then ...... into the bloodstream.
Enzymes ...... up the rate of digestion. The three types of enzyme are ......, ..., and ......

Protease enzymes break down protein in ...... conditions. Bile helps in the digestion of ......
Most food is absorbed in the ....

2 Write down a food that contains a lot of:

a) protein

b) carbohydrate

c) fat

d) minerals and vitamins.

3 Why do marathon runners eat a lot of pasta in the days before a race?

4 Explain why food must be digested.

5 Find out what indigestion tablets contain. Which bit of the digestive system do they work on?

6 What important 'food' is absorbed by the large intestine?

7 What do we call the semi-solid waste that remains?

8 Find out about 'Heartburn'. What is its connection with digestion?

9 This drawing shows part of the small intestine:

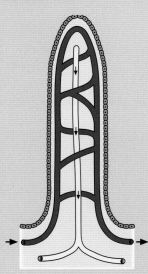

a) What is it called?

b) Write down three features of the small intestine that make it good at absorbing food.

10 An enzyme called Pectinase is used to help make fruit juice. Pectinase breaks down a substance called pectin. Pectin makes fruit juice cloudy.

Explain why pectin is used in the manufacture of fruit juice.

11 Look at this drawing of the digestive system:

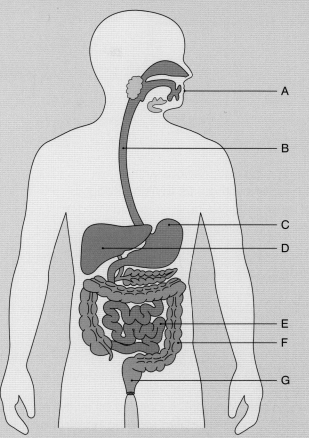

a) Write down the names of the parts A–G.

b) Which part:

    i)   contains acid

    ii) is a muscular tube

    iii) absorbs water

    iv) makes bile

    v)   absorbs most food

    vi) makes an alkali?

12 Enzymes are destroyed by temperatures above about 45°C.

Explain why the instructions on Biological washing powders say: 'use a low temperature wash only'.

CHAPTER **3**

Breathing

▶▶▶ 3a What is breathing?

Breathing is an unconscious activity.
It just happens without us thinking about it.
In fact as soon as we do start to think about
our breathing it changes.

a) How many times do you breathe in one minute?

Breathing is necessary to get a useful gas into your body.
It also removes a waste gas.

b) Can you remember the names of these two gases?

The useful gas, (oxygen), is needed by all the body's cells.
It is used in a process called **aerobic respiration**.

> **Aerobic respiration** is the name for the chemical reaction that releases energy from food.
>
> We can summarise it with this simple word equation:
>
> **Glucose + oxygen → carbon dioxide + water + energy**

We saw in Chapter 2 that carbohydrates are energy-giving foods.
Glucose is an example of a carbohydrate that is a good store
of energy.

c) Can you name any other foods that are good for giving energy?

This reaction is like burning – without the flames!
Some of the energy that is made is released as heat.
This helps to keep you warm.

Carbon dioxide is a waste gas. It is not needed by the body.

d) What living things *do* need this gas?

The water is released as vapour. It is not the water you release
in sweat or urine.

Sometimes breathing is slow.

. . . and sometimes it is fast.

You can see your breath on a cold morning.

Why do we need energy?

If you ask a friend why we need energy,
they usually say 'to keep us alive'.
Well that's true but it's not a very scientific answer.

The energy released by respiration has three main uses:
- It allows us to move by providing energy for muscle contraction.
  Each muscle is made of thousands of individual cells.
  Each cell needs a supply of glucose and oxygen so that it can transfer energy.
  It is not just the muscles of your arms and legs that need energy. Your heart is a muscle and it needs energy to keep beating.

Energy for movement.

e) Can you think of any other muscles that work all the time (just like the heart)?

- Energy is also needed to help us to grow.
  Your body has to build large molecules
  from small ones, and this takes a lot of energy.

Energy for growth.

f) Proteins are large molecules used in growth.
What are the small molecules they are made of called?

- Energy is also needed to help us keep a
  steady body temperature.
  Respiration releases heat energy and this helps
  to keep us warm, even on very cold days.

g) What is the normal body temperature of a human?

Energy to keep a steady temperature.

Remind yourself!

1 Copy and complete:

Breathing gets ...... into the body, and removes
.... Oxygen is used in .... to
release the ...... in ......

The body needs energy for warmth, ...
and ......

2 Explain why respiration is very similar to burning.

3 Find out:

a) what other gases are found in the atmosphere?

b) are any of them useful, and if so how?

The breathing system is in fact a collection of tubes.
It is contained in the chest or **thorax**.
Its job is to get air into and out of the body.
From this air, oxygen will pass into the blood.
At the same time carbon dioxide will pass out.

Exam hint:
Remember breathing and respiration are **not** the same thing.

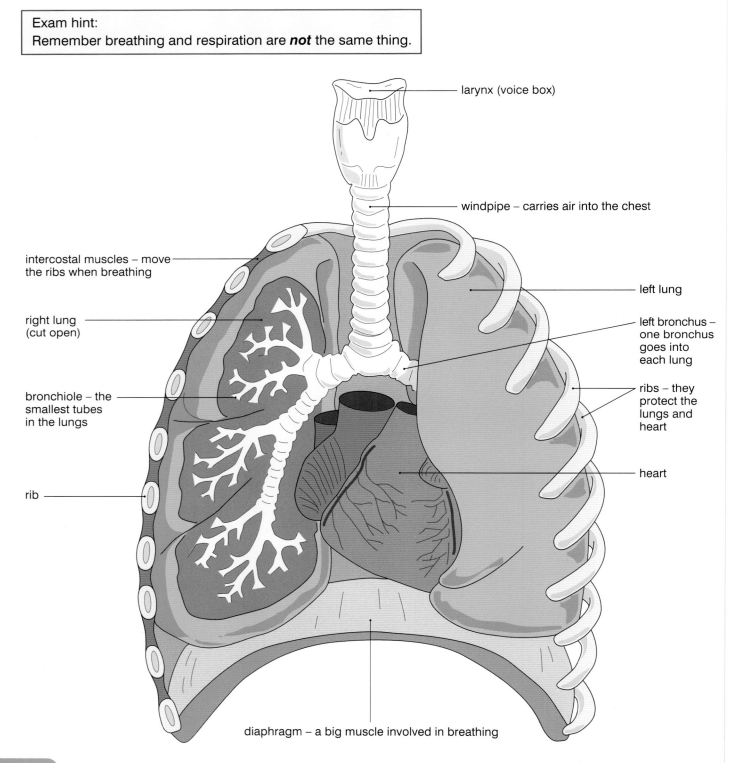

larynx (voice box)

windpipe – carries air into the chest

intercostal muscles – move the ribs when breathing

right lung (cut open)

bronchiole – the smallest tubes in the lungs

rib

left lung

left bronchus – one bronchus goes into each lung

ribs – they protect the lungs and heart

heart

diaphragm – a big muscle involved in breathing

How do we breathe?

- Put your hand on your chest.
  Now take a big breath in.

a) In which direction(s) does your chest move?

The muscles in between your ribs (the **intercostals**) contract.
This pulls your rib cage upwards and outwards.

At the same time your **diaphragm** also contracts
and becomes flat.

These two movements make the space
inside your chest bigger.

As a result, air is drawn in to make your
lungs fill the space.
This is **breathing in**.

- Now put your hand on your chest again.
  This time breathe out as much as you can.

b) In which direction(s) does your chest move this time?

This time your intercostal muscles have relaxed.
This makes your ribs move downwards and inwards.

Also your diaphragm has relaxed, and become
dome shaped again.

These two movements make the space
inside your chest smaller.

As a result, air is forced out of your lungs.
This is **breathing out**.

Breathing in and out is called **ventilation**.

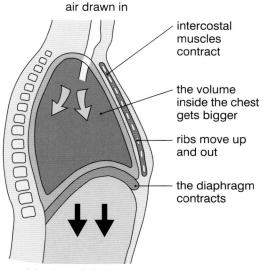

air drawn in

- intercostal muscles contract
- the volume inside the chest gets bigger
- ribs move up and out
- the diaphragm contracts

side view of chest

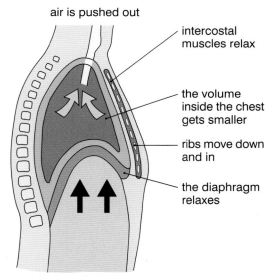

air is pushed out

- intercostal muscles relax
- the volume inside the chest gets smaller
- ribs move down and in
- the diaphragm relaxes

side view of chest

Remind yourself!

1 Copy and complete:

The lungs are found in the ......

They are protected by the ......

When we breathe in the ribs move ...... This
helps to make the volume inside the chest ......

2 The diaphragm separates which two parts of the
body?

3 How does the shape of the diaphragm change
when you breathe out?

What do you think the inside of your lungs are like?
Two big balloons is the usual answer.
In fact if you said like a sponge or a chocolate Aero bar,
you'd be nearer the mark.

The lungs contain millions of tiny bubble-like structures.
These are air sacs called **alveoli**.
Their job is to allow gases to be exchanged between the air
and the blood.

These alveoli are well designed for gas exchange:
- they have **thin walls**
- they are surrounded by **blood vessels**
- they are **moist** (so gases can dissolve) and,
- they have a **massive surface area**.

Imagine all the air sacs in your lungs were flattened out.
They would cover an area as big as a tennis court!

a) By what process do gases pass into and out of
your blood? (Hint: look back at Chapter 1.)

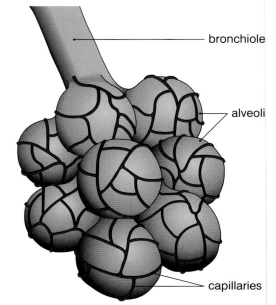

A group of alveoli.

Changing the air

The air we breathe out is different from the air we breathe in.

Look at this table.

Gas	Air we breathe in	Air we breathe out
Oxygen	21%	16%
Carbon dioxide	0.04%	4%
Nitrogen	79%	79%
Water vapour	Varies	Saturated

b) Which gas is produced?

c) Why do we breathe out less oxygen than we take in?

The amount of nitrogen stays the same.

d) What does this tell you about our bodies' use of nitrogen?

Notice that there is a lot more carbon dioxide in the air you breathe out.
This can be very useful.
When someone has stopped breathing, we can give them 'mouth to mouth' (artificial respiration).
This involves breathing out into their lungs.
The extra carbon dioxide they receive often helps to restart their breathing.

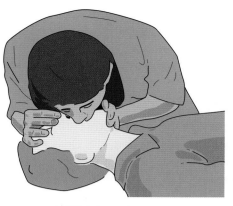
Artificial respiration.

Treating the air

Your chest wall is designed to protect your lungs.

> **e)** How does it do this?

Apart from protecting your lungs from physical injury, your breathing system cleans up the air.

The inside of your nose contains lots of tiny blood vessels. As air passes through your nose, it is warmed up by the blood.

Also, your nose is moist and this makes dry air damp. Warm, moist air is less of a shock to your lungs than dry cold air.

Cilia keep your lungs clean.

Inside your windpipe there are millions of tiny hairs called **cilia** and a lot of sticky **mucus**. Particles of dirt get stuck in the mucus. The hairs then beat backwards and forwards moving the dirty mucus up to the throat.

Smoking damages these hairs and stops them working. This is why heavy smokers have a 'smoker's cough'. This is caused when dirt builds up and irritates their air sacs.
This can lead to diseases like emphysema and lung cancer.

> **f)** Explain how we remove dirt from the air that we breathe in.
>
> **g)** How does smoking damage your respiratory system?

Remind yourself!

1 Copy and complete:

The lungs contain tiny …… called alveoli. These are very …… and they cover a …… area. Gases pass into and out of the blood by …… Before the gases enter the blood they …… in the moisture lining the alveoli.

2 Explain why we breathe out more water than we breathe in.

3 Find out about the disease emphysema. How does it affect the air sacs?

Have you ever tried sprinting **and** breathing deeply
at the same time?
It's not easy!

What this means is that your body cannot get enough
oxygen for aerobic respiration.
Fortunately you can still release energy in muscles even
without oxygen.
This is called **anaerobic respiration**.

Anaerobic respiration can be summarised like this:

Glucose → lactic acid + energy

---

a) Apart from the lack of oxygen, write down two other differences
between aerobic and anaerobic respiration.

Anaerobic respiration is good for releasing energy quickly.
But it has two big problems:
- your cells can only respire like this for a short time
- it produces a lot less energy than aerobic respiration.

As a result you can only do very vigorous exercise such as
sprinting for a short time.

The main waste product in anaerobic respiration is **lactic acid**.
This chemical builds up in your muscles, making them ache.
Eventually it can give a sharp pain, which we usually call a stitch.

Less than 10 seconds later!

The only way to get rid of this pain is to stop exercising
and breathe deeply.
This gets extra oxygen to the muscle cells.
This oxygen is then used to break down the lactic acid
into carbon dioxide and water.

Taking in this extra oxygen is known as paying
the **oxygen debt**.

b) Why do your muscles start to ache
when you are doing a lot of exercise?

c) How can you get rid of a stitch
when doing the school cross country?

Here's the last instalment on my oxygen debt

Summary

The breathing system gets gases into and out of the body. Breathing is also known as **ventilation**.

From this air, oxygen **diffuses** into the blood, and carbon dioxide diffuses out.

The **lungs** are found in the **thorax** and are protected by the **ribs**.

When you breathe in your **ribs** move **outwards** and your **diaphragm** becomes **flat**. This makes the space in your chest **bigger** and air rushes **in**.

Aerobic respiration looks like this:

    glucose + oxygen → carbon dioxide + water + energy

If there is a shortage of oxygen, we respire anaerobically.
Anaerobic respiration looks like this:

    glucose → lactic acid + (less) energy

To get rid of lactic acid we need to take in extra oxygen.

Energy from respiration is used to:
- keep us warm
- make muscles contract
- build large molecules from small ones (helping us to grow).

Questions

1 Copy and complete:

Breathing helps to get …… into your bloodstream. This is needed for …… This process releases ……

A waste gas called …… …… is produced. We breathe out this gas, together with …… …… Without energy our …… would not contract and we would struggle to keep …… Respiration can happen without ……, but this makes much less …… This process is called …… ……

Gases pass into and out of the blood by ……

2 a) Explain the difference between breathing and respiration.

b) What is the difference between gas exchange and breathing?

3 What is your normal breathing rate in breaths per minute?

4 Explain why you can see your breath on a cold morning.

5 Marathon runners can keep going for hours, but sprinters only run for seconds.

Explain this difference.

CHAPTER 4

TRANSPORT

▶▶▶ 4a The circulatory system

Have you been shopping lately?
In shops you will notice how the shelves
are always well stocked.
This is because there is a good transport system.
Lorries and trains keep the shops well supplied with
things to sell.

The human body also has a good transport system.
It is called the **circulatory system**.

It keeps all of your cells supplied with the things
they need. It also removes their waste products.

Shops depend on a good transport system.

a) Write down two things that all cells need for respiration.
(Hint: look at page 26.)

The circulatory system is made up of three main parts:
blood, **blood vessels** and the **heart**.

Look at the drawing opposite.
Can you see that we have two circulatory systems?

● One system takes blood from the heart to the
lungs. This is where gases are exchanged
between the air and the blood.

b) Which gas enters the blood from the lungs?
(Hint: look at page 28.)

● When the blood returns to the heart it is then
pumped all the way around the body.
This carries oxygen and food to the cells.
On this journey it also picks up carbon dioxide
and other waste.

Believe it or not, the whole journey around both
systems takes less than one minute.

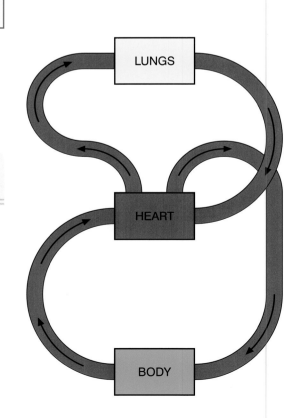

Blood vessels

The tubes that make up most of the circulatory system are called **vessels**.

There are three types of blood vessel. Do you know their names?

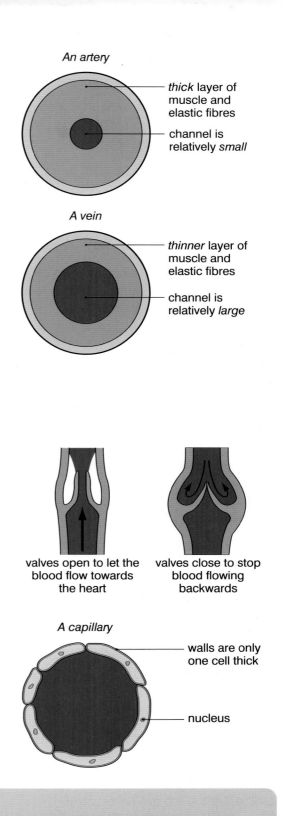

An artery

thick layer of muscle and elastic fibres

channel is relatively *small*

> **Arteries** carry blood *away* from the heart.

- They have thick strong walls. These walls contain lots of muscle and elastic fibres.

Blood flowing in arteries is under high pressure. This is why they need to be strong.

> **Veins** carry blood *towards* the heart.

- Their walls are much thinner. This is because the blood in them is at a lower pressure.

A vein

thinner layer of muscle and elastic fibres

channel is relatively *large*

c) In which type of vessel can we feel our pulse?

Veins have special structures called **valves**.
A valve makes sure that a liquid only flows in one direction.
It can be difficult for the blood to travel back to the heart.
This is because of the low pressure in the veins.
The valves stop the blood flowing back down towards your feet.

valves open to let the blood flow towards the heart

valves close to stop blood flowing backwards

> **Capillaries** are tiny blood vessels. They split off from arteries and eventually rejoin to make veins.

- Their walls are very thin. This allows useful substances to pass into the cells. Waste substances can also then pass out of the cells.

A capillary

walls are only one cell thick

nucleus

Remind yourself!

Copy and complete:

1 Humans have ...... circulatory systems. One collects ...... from the ...... The other delivers oxygen and ...... to the cells. Arteries carry blood ...... from the heart and ...... carry blood towards the heart.

2 Explain why the walls of the capillaries need to be very thin.

3 What word describes how substances pass into and out of the blood?

You've probably heard of heart transplants.
But did you know that we have also
tried mechanical pumps?

> The heart is the pump that keeps the blood
> flowing around your body.

It is made of a special type of muscle.

It is special because unlike your leg muscles
it never gets tired. This is why it can keep
beating all day, every day.

a) Work out your heart rate in beats per minute.

Two pumps in one

On page 34 we saw that there are two circulatory systems.
The heart is split into two pumps side by side.

- The right side of the heart pumps blood
  up to the lungs.
- The left side of the heart pumps blood all
  the way around the body.

The two sides of the heart are completely separate.

- The right side contains blood that has
  returned from the body cells. It has little
  oxygen left in it.
  This is called **deoxygenated** blood.

b) Where will it get new supplies of oxygen
from? (Hint: look at page 28.)

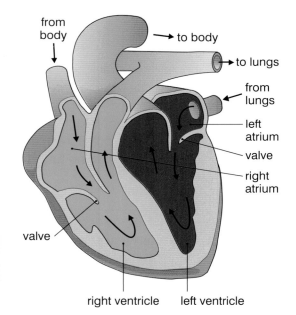

- The left side contains **oxygenated** blood.
  This is blood full of oxygen ready to be
  delivered to the cells.

The heart is also split into four chambers.
The two at the top are called **atria**.
These receive blood from the veins and
then they contract. This squeezes the blood
into the bottom two chambers.
These are called **ventricles**.

Two pumps – one action

To keep the blood flowing smoothly,
both sides of the heart must work together.

1   First the top two chambers fill with blood.

c) What are the top chambers called?

2   When they are full, both of them contract together.
This pushes all the blood into the bottom two
chambers.

d) What are the bottom chambers called?

3   Now these two chambers contract together.
This action forces the blood out of the heart.
It goes into two arteries; one going to the lungs
and one to the rest of the body.

Look again at the drawing of the heart.
Why is it that when the ventricles contract, the blood
doesn't go straight back into the atria?

This is because the heart also contains valves.

When the ventricles contract, the pressure closes
the heart valves. This means that all the blood is
forced to pass out of the heart.

The ventricles of the heart have thicker walls
than the atria.
Also, the left ventricle muscle is thicker than the
right ventricle.

e) Why does the left ventricle need a stronger muscle?

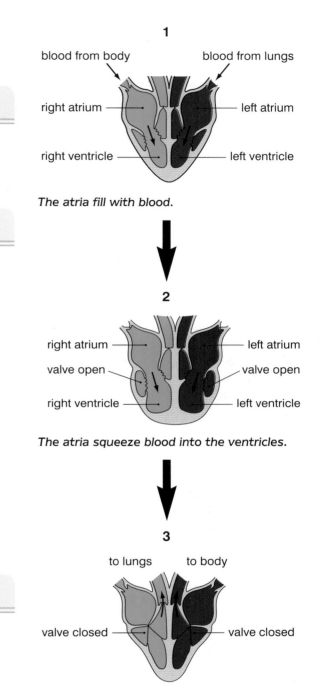

1

blood from body   blood from lungs

right atrium   left atrium

right ventricle   left ventricle

The atria fill with blood.

2

right atrium   left atrium

valve open   valve open

right ventricle   left ventricle

The atria squeeze blood into the ventricles.

3

to lungs   to body

valve closed   valve closed

The ventricles now squeeze blood out of the heart.

Remind yourself!

Copy and complete:

1 The heart is a ......, made of ...... The ...... side
deals with blood that is ...... of oxygen. The top
chambers are called ......, and the bottom are
called ......

2 Why are valves needed inside the heart?

3 The normal heart rate is around 70 beats per
min. Name two situations when this is likely to
increase.

Blood is the liquid that carries substances around the body.

There are around 5 litres of blood in your body.
New blood cells are being made all the time.
This is because old cells die.
New cells are mainly made in the **bone marrow**.
This is the soft centre of bones that dogs like to eat.

If you have an accident or a big operation then
you will need a **blood transfusion**.
This is where you are given blood to replace
what you have lost.

You cannot just be given any blood.
Each person has one of four blood groups called:
A, **B**, **AB** or **O**.
Being given the wrong group could be fatal.

a) Do you know what your blood group is?

Blood is made up of **red cells**, **white cells**,
and bits of cells called **platelets**. All of these
are carried in a liquid called **plasma**.

Red cells

Red cells contain a chemical that gives blood its colour.
There are 5 000 000 red cells in 1 cm^3 of blood.
Red cells are shaped like Polo mints but with a dent
instead of a hole.

Look at this drawing of a red cell.

b) What is missing that you would normally expect to see
in an animal cell?

The job of red cells is to carry oxygen from the lungs to all the cells
of the body.

Not having a nucleus means that there is more room for oxygen.
Their special shape also gives them a big surface area,
so they can pick up lots of oxygen.

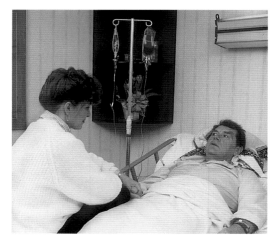

A life saving transfusion.

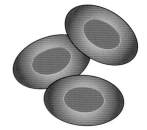

Red blood cells.

A red blood cell cut open.

White cells

There are far fewer white cells than red ones.
White cells, though, are much bigger and they *do* have a nucleus.

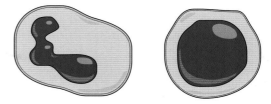

> The job of white cells is to help the body fight off microorganisms (like bacteria and viruses) that cause disease.

White blood cells often have a large or strangely shaped nucleus.

Platelets

Platelets are really just small bits of cells.
Just like red cells they do not have a nucleus.

> Platelets help the blood to clot and stop bleeding at cuts.

The platelets help to make a net of tiny fibres over the cut.
This net then traps red cells, blocking the cut and stopping the bleeding.
This blockage is called a **clot** and it hardens into a scab.
Underneath the scab, new skin grows and eventually the scab drops off.

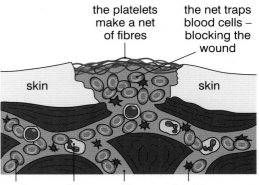

Plasma

Plasma makes up **slightly more than half** of your blood.
It is a straw coloured liquid. It is mainly water plus dissolved substances such as:

- carbon dioxide carried from cells to the lungs
- digested food carried from the gut to the cells
- urea (a waste product) carried from the liver to the kidneys.

Plasma is very important.
If there is a shortage of liquid in your veins and arteries your heart will stop working.
Doctors often give patients plasma while they are trying to find out their blood group.

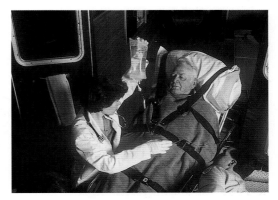

Plasma is a life saving fluid.

c) What is plasma?

Remind yourself!

Copy and complete:

1 Blood is made up of a liquid called …… It also contains …… and …… cells. Blood helps to …… substances around the body, and …… disease.

   …… help to prevent bleeding from cuts.

2 Explain why red cells are suited to carrying oxygen.

3 How do scabs over cuts help to stop wounds becoming infected?

4 Draw a mind map to show all you know about blood.

Summary

The circulatory system transports substances around the body.

Some substances like dissolved food and oxygen are useful. Others like carbon dioxide are waste products.

Arteries carry blood **away** from the heart and **veins** carry blood **towards** the heart.

Arteries have **thick muscular walls** but veins are much thinner.

Veins have **valves** to stop the backflow of blood.

Capillaries have **very thin walls** that let substances **diffuse** through.

Humans have a **double circulatory system**. One system takes blood to the lungs to pick up oxygen. The other system takes this blood to all the cells of the body.

Blood is pumped by the **heart**, which is made of muscle. The **atria** contract forcing blood into the **ventricles**. The ventricles contract forcing blood out of the heart.

Blood is made up of **plasma** (a liquid), **red cells**, **white cells** and **platelets**. Plasma carries dissolved substances like carbon dioxide and food. Red cells carry oxygen, white cells fight disease and platelets help to seal cuts.

Questions

1 Copy and complete:

The circulatory system needs a pump called the ...... This is mainly made of ...... Veins bring blood ...... the heart, while ...... take blood away from it. The heart contains ...... to keep the blood flowing the right way. The liquid part of blood is called ...... It carries ...... substances like carbon dioxide. Red cells carry ...... gas.

White cells are ...... than red cells. They help the body to fight ......

2 Copy and complete this table:

Arteries	Capillaries	Veins
Have ...... walls	Have very ......walls	Have ...... walls
Carry blood ...... from the heart	Join up ...... and	Carry blood ...... the heart.
Do not have	Do not have	Have ......
Have a pulse	No	No ......

3 Look at this diagram of the human heart.

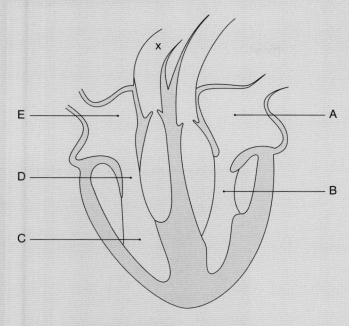

i) Name the parts labelled A to E.

ii) Why is the left ventricle thicker than the right?

iii) Vessel x is leaving the right side of the heart.
Where is it carrying blood to?

4 i) Draw a red blood cell and a white blood cell to show two differences between them.

ii) What is the main difference between these cells and platelets?

5 At high altitude there is less oxygen in the air. People who live at altitude have more red cells in their blood.
This table compares the red cells in the blood of 3 people:

	Person 1	Person 2	Person 3
Number of red cells (mm^3)	8,000,000	5,000,000	2,000,000

a) Which person lives at the highest altitude?

b) Explain your answer.

c) Why is it an advantage to have more red cells at high altitude?

6 Do some research on heart disease and answer these questions:

i) What types of people are most at risk from heart disease?

ii) What sort of lifestyle should you lead to avoid heart disease?

iii) Design a poster summarising this advice.

7 Draw a capillary and clearly show:

i) How it is designed to let substances through easily.

ii) What substances enter from the cells.

iii) What substances leave from the cells into the blood.

8 Draw a simple diagram to show what is meant by a double circulatory system.

9 Do some research and find out about the circulatory system in a fish.
How is this different to humans?

CHAPTER 5

▶▶▶ 5a Microbes and disease

Most diseases are caused by tiny living things called **microbes**. You can tell by their name that they can only be seen with a microscope.

In the summer of 2001 a disease called 'foot and mouth' affected farms all over Britain. This disease led to hundreds of thousands of cows and sheep being slaughtered.

This disease is caused by one type of microbe called a **virus**. The government decided that animals with the disease must be killed and burned. This was to try to kill off the virus.

a) Why were animals on nearby farms also killed?

b) Can you name any human diseases caused by a virus?

Huge fires were needed to burn infected animals.

The other main type of microbe that cause disease are called **bacteria**.

c) Can you name any human diseases caused by bacteria?

Other microbes

There are also two other kinds of microbe:

- **fungi** – not large mushrooms, but microscopic cells. Certain fungi can cause disease. Athlete's foot for example happens when tiny fungal cells grow in between your toes.
- **Single-celled organisms**. Malaria for example, is caused when these tiny organisms are injected into your blood by mosquitoes.

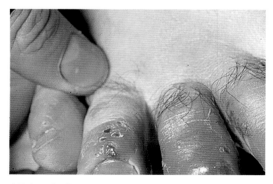

Athlete's foot.

Bacteria and viruses

Bacteria are similar to animal cells. They have a cell membrane and cytoplasm. But there are also three big differences:
- they have **no** nucleus
- they **do** have a cell wall outside the cell membrane
- they are much smaller.

round   spiral shaped   rod shaped

Bacteria come in a variety of shapes.

d) Plant cells also have a cell wall (although it is a different type). What does a cell wall do?

Viruses are not really like cells at all. They are made up of a protein coat which surrounds a few genes.

Viruses can be very strange looking – not like cells at all!

Genes are the structures that hold all the information about the design of living things. Unlike bacteria, viruses can only live and reproduce inside other living cells. Viruses are also much smaller than bacteria.

Both bacteria and viruses can reproduce very quickly. For example, bacteria can reproduce as often as once every 20 minutes if the conditions are right.

How do bacteria and viruses affect us?

There are two ways that these microbes can make us feel ill.
- They can cause damage to living tissue (e.g. the flu virus will damage the cells in the nose and throat).
- They can also produce **toxins** (poisons), which make us feel ill. For example, the bacteria that give us food poisoning make plenty of these toxins.

Remind yourself!

Copy and complete:

1 Microscopic living things are called ...... Some of them can cause ......

The two main disease-causing microbes are ...... and ...... Both are dangerous because they ...... very quickly.

2 Why should we always cook meat thoroughly?

3 Draw a table to show three differences between bacteria and viruses.

In large communities like schools, if one person has flu then lots of people tend to catch it. This is because diseases spread easily where there are large numbers of people together.

Disease also spreads easily in unhygienic conditions. People in poorer countries often have to drink dirty water. This water may contain thousands of bacteria and so disease will spread very quickly.

Fortunately our bodies have a number of ways of stopping microbes from entering:

Coughs and sneezes spread diseases!

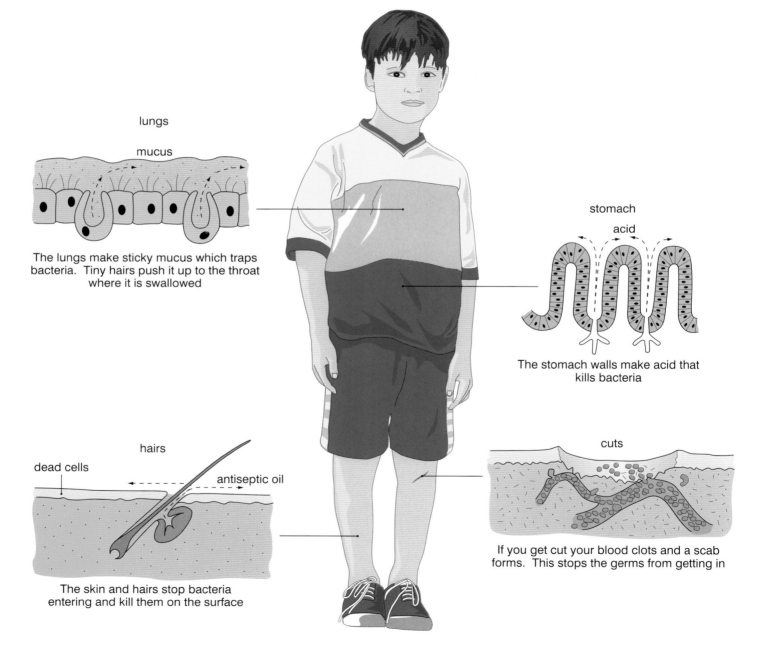

lungs

mucus

The lungs make sticky mucus which traps bacteria. Tiny hairs push it up to the throat where it is swallowed

stomach

acid

The stomach walls make acid that kills bacteria

hairs

dead cells

antiseptic oil

The skin and hairs stop bacteria entering and kill them on the surface

cuts

If you get cut your blood clots and a scab forms. This stops the germs from getting in

The blood fights back!

If disease-causing microbes have entered our body we can rely on the blood to help out.
White blood cells can be produced in large numbers to help kill microbes.

Have you ever been told that 'your glands are up'? When you are ill, glands in your neck produce lots of white blood cells. As a result they become swollen and painful.

a) What is another common symptom of being ill?

White blood cells help to fight infection.

They do this in a number of ways:

● By **ingesting** (eating) the microbes. They can do this because their cytoplasm can flow enabling them to surround bacteria. Then they destroy the bacteria with powerful enzymes.

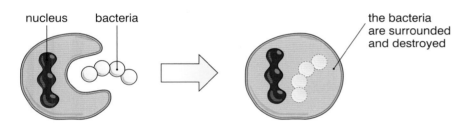

nucleus   bacteria

the bacteria are surrounded and destroyed

● By making special chemicals called **antibodies**. These chemicals stick onto the microbes, making it easier for them to be destroyed. They can also remain in the blood for a long time. This gives us long-term protection.
● By making other chemicals called **antitoxins**. These counteract the effects of the toxins (poisons) made by the microbes.

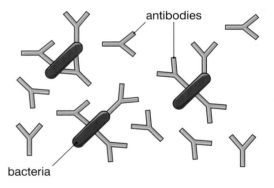

antibodies

bacteria

Remind yourself!

Copy and complete:

1 Unhygienic and …… conditions are ideal for the spread of …… Our …… is very good at keeping microbes out. If microbes do enter, our …… blood cells are the next line of defence.

2 Briefly describe the three ways that white blood cells fight infection.

3 Find out what diseases are commonly spread through dirty water.

Why are these diseases not common in Britain?

As you probably know, it is quite common to receive a vaccination in school. This is when everybody lines up outside the hall with their sleeves rolled up. Your name is called and you walk in to receive the dreaded needle!

a) What is the name of the last vaccination you had?

A **vaccination** is an injection of dead or weakened bacteria. This injection is called a **vaccine** and it gives you a mild form of a disease.

A very common vaccination is against the bacteria that can cause tetanus (often called lock jaw).
Although the bacteria are weak, they can be recognised by your white blood cells. These cells then make antibodies against the bacteria.
These antibodies will remain in your blood for months or even years.
The next time you catch the same bacteria these antibodies will be ready to deal with them. This means that although the bacteria enter your body, you will not suffer any **symptoms**.

Vaccinations can also be given against viruses, such as influenza (flu), or measles.

After you have had a vaccination, you might feel a bit unwell. This is because even though the microbes are weak they can still cause symptoms before they are destroyed.

Have you ever been on an exotic holiday?
If you go abroad to countries such as Africa then vaccinations are usually needed.
These are to protect you against diseases like cholera that are not common in Britain.

After a vaccination you are said to be **immune** to the disease.

b) Try to explain why you rarely catch diseases like measles more than once?
(Even without the benefit of a vaccination.)

Summary

Diseases are often caused by microbes that enter the body.
The most common microbes that cause disease are **bacteria** and **viruses**.

Bacteria are cells that have cytoplasm, a membrane and a wall but **no** nucleus.
Viruses are much smaller and they simply have a protein coat and a few genes.

Microbes can reproduce very quickly.
They can also spread quickly if conditions are unhygienic or overcrowded.

The body is very good at preventing the entry of microbes.
The skin, breathing system and blood act as barriers to microbes.

White blood cells can destroy microbes or make **antibodies** against them.
They can also make **antitoxins**, which counteract the poisons produced by the microbes.
These toxins often cause the symptoms of a disease.

Vaccinations protect people against disease.
They contain weakened microbes, which give you a mild form of a disease.
These weak microbes cause your white cells to make antibodies against the disease.
These antibodies can remain in your blood for years.
They give you **immunity** against a disease.

Questions

Copy and complete:

1 Bacteria and ...... are both microbes that cause ...... These microbes can ...... very quickly in ...... conditions. ...... blood cells are very good at fighting ...... They can destroy microbes by ...... them. They also make ...... These can stay in the blood for a ...... time. The symptoms of a disease are often caused by the ...... the microbes release. Vaccinations can be used to give ...... to a disease. These contain ...... microbes.

2 These terms are all associated with disease, find out what they mean:

   a) symptoms

   b) incubation period

   c) infectious

   d) contagious.

3 Describe with the aid of drawings how white blood cells ingest microbes.
(Hint: look on page 45.)

4 Explain why schools are ideal places for the spread of diseases like flu.

5 Explain how your stomach stops bacteria getting into your blood. (Hint: look on page 44.)

6 Find out about Edward Jenner and Jonas Salk. What were their contributions to the development of vaccinations?

7 Find out which diseases the MMR vaccine gives protection against.

▶ **Food and digestion**

1 The diagram shows the human digestive system.

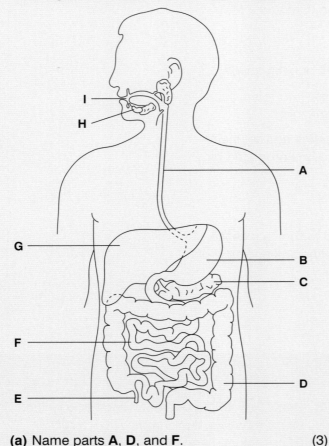

(a) Name parts **A**, **D**, and **F**. (3)

(b) Describe **two** things that happen to food in the digestive system. (2)

(c) Use the letters from the diagram to show where:

(i) hydrochloric acid is produced (1)

(ii) water is absorbed (1)

(iii) carbohydrases are produced. (1)

(d) What do carbohydrases do in the digestive system? (2)

(AQA (NEAB) 1999)

2 (a) Milk contains carbohydrate, protein and fat. Give **one** reason why the human body needs:

(i) carbohydrates; (1)

(ii) proteins; (1)

(iii) fats. (1)

(b) Milk does not contain fibre. Why is it important that a person's diet contains some fibre? (2)

(c) The table shows some of the nutrients contained in cows' milk and in soya milk. Soya milk is made from vegetable products.

Nutrient	Mass contained in 100 g of milk (g)	
	Cows' milk	Soya milk
Carbohydrate	4.8	5.0
Protein	3.1	3.7
Saturated fat	2.5	0.2
Unsaturated fat	1.4	1.5

People may be advised, for health reasons, to drink soya milk instead of cows' milk. Explain why soya milk is more healthy than cows' milk. (3)

(AQA 2001)

3 This table shows what happens to the food that you eat.

Part of body	Time spent there by food
mouth	a few seconds
gullet	a few seconds
stomach	2–4 hours
intestines	10–20 hours

(a) Choose words from this list to complete the sentences below.

anus   blood   small intestine   stomach

The food you eat is broken down into soluble substances.
These substances are then absorbed through the walls of your ...... so that they can pass into your .... Undigested food (faeces) passes out of your body through your .... (3)

(b) (i) In which part of your digestive system does food spend the longest time? Choose from the list below.

mouth   gullet   intestines   stomach

(ii) How long does it take for food to pass all the way through your digestive system? (2)

(AQA 2001)

4 The table shows the amounts of carbohydrate, fat and protein in 100 g portions of five foods, A–E.

Food	Mass in 100 g portion (g)		
	Carbohydrate	Fat	Protein
A	0	1	20
B	50	2	8
C	0	82	0
D	12	0	1
E	20	0	2

(a) Which food:

    (i)   contains most carbohydrate;

    (ii) is butter;

    (iii) is best for replacing cells?   (3)

(b) A person eats 50 g of food E. How much carbohydrate would the person eat?   (1)

(c) Describe, in as much detail as you can, what happens to the protein after food A is swallowed.   (4)

(AQA (NEAB) 2000)

5 Some of the food we eat is digested and then absorbed into our blood.
The rest of the food is passed out of our bodies as faeces.
The pie-charts show what happens to the food in two different meals.

Meal A   • beans on toast
                   • apple

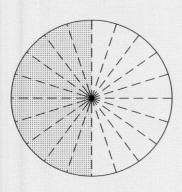

Meal B   • sausages and chips
                   • chocolate bar

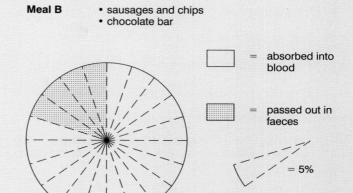

☐ = absorbed into blood

▨ = passed out in faeces

◸ = 5%

[We also absorb most of the water in food. The pie-charts don't show this.]

(a) Copy and complete the table.

	Percentage (%) absorbed	Percentage (%) in faeces
Meal A		
Meal B		

(3)

(b) Suggest a reason for the difference in figures.   (1)

(AQA (NEAB) 1998)

▶ **Breathing**

6 This question is about cells using energy.

The diagram shows a working muscle cell.

The cell releases energy from a chemical reaction.

This reaction takes place inside the cell.

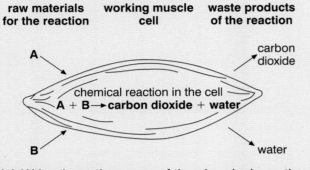

raw materials for the reaction   **working muscle cell**   **waste products of the reaction**

A
carbon dioxide

chemical reaction in the cell
A + B → carbon dioxide + water

B
water

(a) Write down the name of the chemical reaction which releases the energy. (1)

(b) Look at the diagram.
Write down the names of raw materials **A** and **B**. (2)

(c) Describe what happens to the carbon dioxide made in the muscle cell. (2)

(OCR 1999)

7 The diagram shows part of the breathing system in a human.

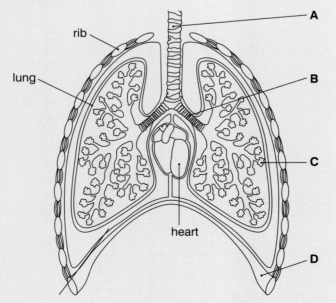

rib
lung
A
B
C
heart
D

(a) Use words from the list to name parts **A–D**.

alveoli   bronchiole   bronchus
diaphragm   trachea (windpipe)   (4)

(b) Where in the lungs does oxygen enter the blood? (1)

(AQA (NEAB) 1999)

8 **List A** gives the names of five parts of the breathing system.
List B gives information about each of these parts in a different order.

List A	List B
alveolus	protects the lungs
bronchus	carries air through the neck
diaphragm	where oxygen passes into the blood
ribcage	divides into smaller branches
trachea	separates the lungs from the abdomen

Copy these lists.
Draw a straight line from each part in **List A** to the correct information in **List B**. One has been done for you. (4)

(AQA (NEAB) 1999)

▶ **Transport**

9 **(a)** What is the job of the heart? (2)

(b) The diagram shows a section through the heart.

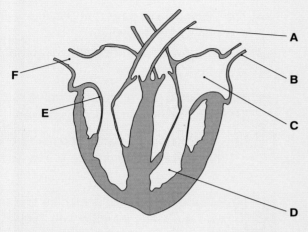

Use words from the list to name the parts labelled **A** to **D**.

artery	**atrium**	**capillary**
valve	**vein**	**ventricle** (4)

(c) What is the job of part **E**? (2)

(d) The blood in part **B** comes from the lungs. The blood in part **F** comes from the body.
Copy and complete the sentences about the composition of blood in parts **B** and **F**.
The blood in part **B** contains a higher concentration of …… than the blood in part **F**.
The blood in part **B** contains a lower concentration of …… than the blood in part **F**.
(2)
(AQA (NEAB) 2000)

10 Blood is made of many parts.

(i) Copy and complete the table to show the job of each part of blood.
The first one has been done for you.

Part of blood	Job
plasma	carries substances round the body
platelet	
red blood cell	

(2)

(ii) The plasma carries many substances round the body.
Write down the name of **one** of them. (1)
(OCR 1999)

11 The drawings show the structure of three types of blood vessel, **A**, **B** and **C**. They are drawn to the scales indicated.

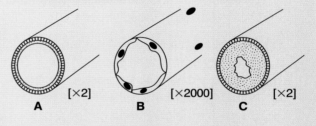

(a) Name the **three** types of blood vessel. (3)

(b) Describe the job of blood vessel **B**. (2)
(AQA (NEAB) 2000)

▶ Disease

12 The diagram shows a section through a virus. This virus multiplies inside human liver cells.

(a) (i)   Name the parts labelled **A** and **B**. (2)

(ii)   Name **two** parts of a liver cell that are **not** found in a virus. (2)

(b) When the virus has multiplied inside the liver cell, the cell bursts releasing viruses into the blood.
Describe how the white blood cells react to the presence of these viruses. (2)

(c) People addicted to heroin often inject themselves with the drug.
This virus is far more common in heroin addicts than in the general population.
Suggest an explanation for this. (2)
(AQA (NEAB) 2000)

13 Read the passage

Could we be on the verge of a modern TB plague?

A dangerous new strain of the bacteria which cause the lung disease, tuberculosis (TB), has recently been found in Britain. Although the victims were treated successfully, doctors fear that the new strain could soon spread. 'All it would take would be for someone infected with the new strain to cough in a place like a cinema and the infection could spread like wildfire,' said a TB specialist. 'No-one is immune to the new strain, and many of the people breathing in the bacteria would develop the disease.'

(a) Tuberculosis is an infection of the lungs.
Most bacteria that we breathe in do not reach the alveoli in the lungs.
Describe how the body prevents bacteria reaching the alveoli. (2)

(b) The TB specialist says that 'No-one is immune to the new strain'.
Explain how we can naturally become immune to a disease. (3)

(c) Many people are immune to the old strain of TB because they have had a vaccine.

(i)   What does a TB vaccine contain that makes a person immune to the disease? (2)

(ii)   It is too late to give a vaccine to a person who is already infected with TB.
What can be injected to stop the disease developing? (1)
(AQA (NEAB) 2000)

14 There are two types of microbe: bacteria and viruses.
When harmful microbes enter our bodies, they can make us ill.

(a) Give three ways that harmful microbes can enter our bodies. (3)

(b) White blood cells help to protect our bodies from bacteria and viruses.
Describe two things that white blood cells can do to protect us from harmful microbes. (4)
(CGP Sample Question)

Section Two
Maintenance of life

In this section you will find out about the systems that
keep humans alive.
You will find out about the systems that keep plants alive.
You will also learn about the effects of drugs on the human body.

CHAPTER 6

Plant life

▶▶▶ 6a Photosynthesis

Light is an important source of energy.

You can buy calculators that are solar powered. There is a row of light sensitive cells above the screen. These cells turn light energy into the electrical energy that runs the calculator.

In hot countries, like Spain, many homes have solar panels on the roof. These absorb sunlight and provide a cheap source of heating.

Light is also an important source of energy for plants.

a) Which part of a plant acts like a solar panel?

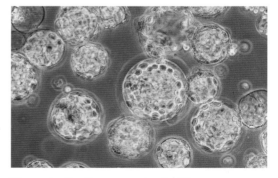

Solar power – a cheap source of energy.

Unlike animals, plants have to make their own food.
To do this they need raw materials and a source of energy.

The raw materials a plant needs are:
- carbon dioxide gas
- water.

b) Where will the plant get water from?

Light provides the energy source, but it cannot be absorbed without a special chemical called **chlorophyll**.

Chlorophyll is green in colour and is found in many plant cells.
It is contained in tiny structures called **chloroplasts**.

> When plants make their own food, it is called **photosynthesis**.
> We can summarise this process like this:
>
> Carbon dioxide + water (+ light energy) → glucose + oxygen

The glucose made is very quickly converted into starch.
Starch is much easier to store because it is a large molecule and it does not dissolve easily.

These leaf cells are packed with chloroplasts.

c) To the plant oxygen is a waste gas, but in what way is it very important to animals?

Using the food

The glucose a plant makes is used in lots of ways.

- Some of it is converted into **starch**. It is then stored for later use. Potatoes and seeds are two good examples of starch stores. When the plant requires energy for growth, the starch is turned back into glucose again.
- Glucose can also be turned into **fats** and **oils**. These can also be stored, for example in linseeds.

> **d)** What do cricketers use linseed oil for?

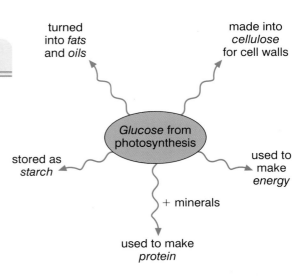

Plant cell walls are very tough. This is because they contain a substance called **cellulose**. Cellulose is another chemical that can be made from glucose.

In Chapter 2 we saw how proteins are needed for growth.
Animals eat protein in their diet but plants can make protein using glucose.
To do this they also need **minerals**.
These are chemicals that plants absorb from the soil.

Plants respire too

Do not forget that plants need to release energy.
They do this in just the same way that we do,
by carrying out respiration.
Some of the glucose made in photosynthesis
is used in this process.

> **e)** Write down the equation for respiration.

Photosynthesis only occurs in the light
but respiration is happening all the time.

Remind yourself!

1 Copy and complete:

Plants make food by a process called ......
Sunlight provides the energy for this process.
The plant also needs .... from the air, and
...... from the soil. ...... gas is produced and is
released into the ......

2 Write down the equations for photosynthesis and respiration one above the other.

What do you notice about them?

3 Why is starch easier to store in cells than sugar?

(Use the ideas of molecule size and solubility in your answer.)

Have you been inside a greenhouse at a garden centre?
If so you will have seen heaters and lamps.
These are not used on long, hot summer days.
But early in the spring when the days are shorter
and colder, they are very useful.

For plants to grow quickly the **rate** of photosynthesis
needs to be as high as possible. The rate could be
measured by looking at how much glucose is made
in a certain time.

We know that glucose is used to help growth.
So if photosynthesis is faster, then more glucose
will be made and the bigger the plant will grow.

Conditions are carefully controlled in here.

> There are three **factors** that affect the rate of photosynthesis:
> * the amount of light
> * the temperature
> * the amount of carbon dioxide.
>
> These are called the **limiting factors**.

Why are they called limiting factors?
Well, whichever one is in short supply **limits**
(slows down) the rate of photosynthesis.

a) In which parts of the world do you think limiting factors
are less important?

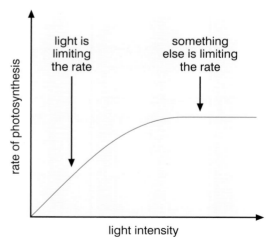

Good conditions for photosynthesis all year round.

Why is light important?

Light is the energy source for photosynthesis.
It makes sense that if there is more light
photosynthesis will go faster.

Look at the graph opposite:

As the amount of light goes up
so does the rate of photosynthesis.
But eventually the graph flattens out
and extra light makes no more difference.

At this point something else is **limiting** the rate.

b) What does 'limiting the rate' mean?

How light affects photosynthesis.

Why is carbon dioxide important?

Carbon dioxide is another limiting factor.
Carbon dioxide is one of the raw materials
for photosynthesis.

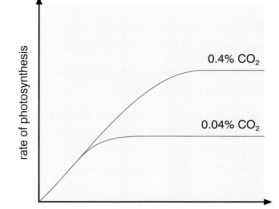

How CO_2 affects photosynthesis.

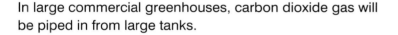

c) What is the other raw material?

If we provide more carbon dioxide,
photosynthesis will be able to go at a faster rate.
Once again this means more food can be made.

There is not very much carbon dioxide in the atmosphere,
only about 0.04%.
In small greenhouses, paraffin heaters are used.
These give off carbon dioxide when the fuel is burnt.

d) What is the other advantage of using these heaters?

In large commercial greenhouses, carbon dioxide gas will
be piped in from large tanks.

Why is temperature important?

Adding extra carbon dioxide will not make any difference
if the temperature is too low.
The ideal temperature for photosynthesis is around 25°C.
Photosynthesis is a chemical reaction and it is controlled
by enzymes.
We saw in Chapter 2 that enzymes are sensitive to
temperature:

- too low and they do not work very well
- too high and they can be destroyed and not work at all.

We can now see that for the best rate of photosynthesis
we need a combination of:

- plenty of light
- a good supply of carbon dioxide
- warm conditions.

Photosynthesis has an optimum (ideal)
temperature.

Remind yourself!

1 Copy and complete:

The ...... of photosynthesis depends on three
...... factors. These are ......, ... and ...... In
...... growers can easily control these factors.

2 What will happen to photosynthesis if the
temperature is too high?

Explain your answer.

3 Explain fully why plants do not grow very well in
the winter.

▶▶▶ 6c Plant food

When you buy a houseplant the label often says 'feed regularly during the growing season'.

But we have already seen how plants make their own food. So what does this mean?

Plants can make glucose by photosynthesis. But they need other chemicals to make things like proteins for growth. These other chemicals are called mineral salts.

a) Where will a plant get these minerals from?

The main minerals needed by plants are:

● **nitrates**
for healthy leaves and good all-round growth

● **phosphates**
for healthy roots

● **potassium**
for healthy flowers and fruit.

These minerals contain the elements **nitrogen (N) phosphorus (P)** and **potassium (K)**. That is why we usually refer to them by their chemical symbols (**NPK**).

Another important element needed by plants is **magnesium**.

b) What is the chemical symbol for magnesium?

Magnesium is needed to make chlorophyll. Without it plants will have yellow leaves that cannot carry out photosynthesis.

Plants that lack minerals will have an unbalanced diet. Therefore, just like humans, they will have deficiency diseases.

Every spring many gardeners add **fertilisers** to their lawn. They do this to encourage the grass to grow after the winter.

c) Which mineral in particular will lawn fertiliser contain?

A plant suffering from a lack of nutrients.

Fertilisers

It is not just gardeners who give their plants minerals. Farmers add huge amounts of fertilisers to their crops. They do this so that they can get the best growth possible.
The more crops they produce, the more money they make.

The elements that farmers add are the same as gardeners.
However farms are much bigger than gardens.
So farmers use machinery to add the chemicals, such as sprays fitted to tractors.

The fertilisers that most farmers and gardeners use are made by humans.
They are often called **artificial fertilisers**.

There is a more natural source of fertiliser though.
Natural fertilisers come mainly from animals.
From animal dung to be precise!
This **manure** is a cheap and readily available fertiliser.

Many farmers prefer artificial fertilisers because:
- they know exactly what elements they contain
- they are easier to transport
- they are easier to put on the land
- there is not enough manure available.

But farmers who produce **organic crops** can *only* use natural fertilisers.
These crops are said to taste better than those grown with artificial fertilisers.
But they are more expensive as they are harder to grow in large quantities.

Manure – smelly but effective.

d) What are organic crops?

e) Why do people buy organic foods even though they are more expensive?

Remind yourself!

1 Copy and complete:

Plants need …… in order to grow properly. The main minerals are …… …… and ……
Magnesium is also important for making ……

2 Which part of a plant will take in the minerals from the soil?

3 Why is magnesium so important to plants?

4 Draw a mind map to summarise the information on these two pages.

Have you ever eaten celery?
If so you will have probably got the
stringy bits caught in your teeth.
These stringy bits are in fact bundles
of transport vessels.

Just like animals, plants also need
a transport system.

---

a) What substances need to be transported around a plant?

Xylem vessels

These tubes carry water and dissolved mineral salts.
Xylem tubes carry these materials from the roots,
up the stem to the leaves.

Phloem vessels

These tubes carry the products of photosynthesis.
Things like glucose (and amino acids)
are made in the leaves.
The phloem vessels carry these materials
to all parts of the plant, in particular, to the growing points (buds).
Also to storage organs, like the potato tuber.

One difference between xylem and phloem
is the direction of transport.
In the xylem, water flows one way only
from the roots to the leaves.
But in the phloem, dissolved food
flows in both directions.

Look at this photograph of a leaf:
Can you see the leaf veins spreading over its surface?
Leaf veins are, in fact, bundles of xylem and phloem vessels.
These bring water right to the leaf cells.

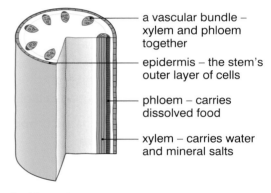

a vascular bundle –
xylem and phloem
together

epidermis – the stem's
outer layer of cells

phloem – carries
dissolved food

xylem – carries water
and mineral salts

Inside a stem.

Roots

Without a good root system, plants cannot grow.
The roots spread out in the soil and absorb water.

b) What else will they absorb with the water?

A good root system also fixes (**anchors**)
a plant firmly in the soil.
This is especially important for big trees.

To help them absorb water, roots also have **root hairs**.
These are tiny hairs that grow on the outside of the root.
They give the root a much bigger **surface area**.
The bigger the surface area, the more water they
can absorb.

How does water enter the roots?

Water passes into the roots by **osmosis**.

> Osmosis is the **diffusion** of water. **Water molecules**
> pass from a **dilute solution** to a **more concentrated**
> one.
> This happens through a **partially permeable**
> **membrane**.

In other words, osmosis is just a special type of diffusion.
It *only* involves the movement of water molecules.

The solution in the soil is always more dilute
than the solution inside the root cells.
So water passes through its root cell membrane
and into the root cells.
Notice that the cell membrane is called *partially*
permeable.
This is because it only lets small molecules (like water) through.

c) What does the word 'permeable' mean?

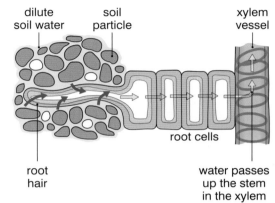

Water enters the root by osmosis.

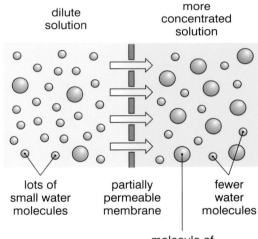

Osmosis is the movement of water molecules.

Remind yourself!

1 Copy and complete:

Plants have …… transport systems. The ……
tubes carry water from the roots to the ……

The …… tubes carry food from the …… to all
parts of the plant.

2 Why are root hairs so important to young
growing plants?

3 Why do the few plants that grow in deserts have
very long root systems?

The leaf is one of the most important parts of a plant.

Most photosynthesis takes place in leaves.

a) Why is the leaf ideally suited to photosynthesis?

Look at this drawing of a section through a leaf:

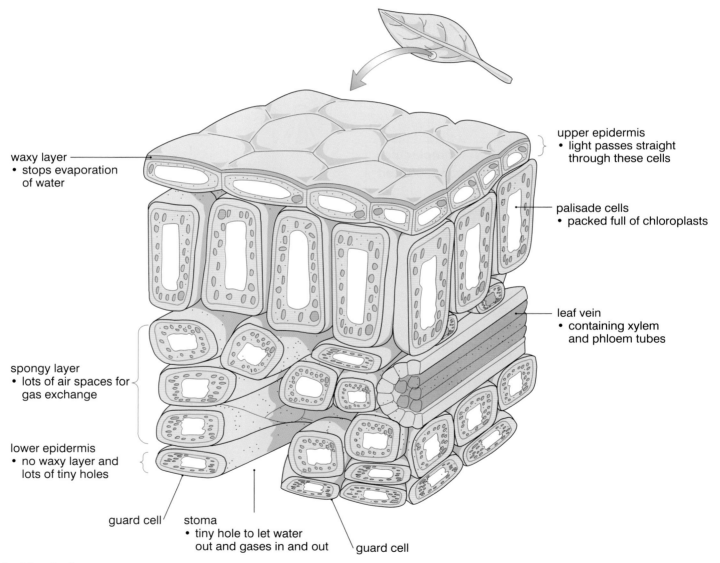

waxy layer
• stops evaporation of water

upper epidermis
• light passes straight through these cells

palisade cells
• packed full of chloroplasts

leaf vein
• containing xylem and phloem tubes

spongy layer
• lots of air spaces for gas exchange

lower epidermis
• no waxy layer and lots of tiny holes

guard cell

stoma
• tiny hole to let water out and gases in and out

guard cell

Inside a leaf.

Transpiration

> **Plants lose water vapour from the surface of their leaves.**
> We call this loss **transpiration**.

The water **diffuses** through tiny holes called **stomata**.
These holes are found mainly on the bottom of the leaf.
Stomata can open and close. This happens
because they are surrounded by sausage-shaped
guard cells.

When the guard cells swell up, they open up the stomata.
This allows water vapour out.

When the stomata are open they also let carbon dioxide
gas in.

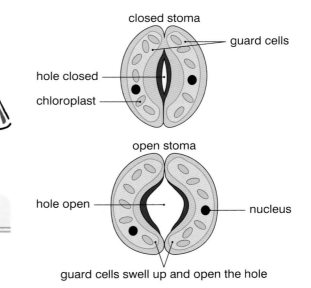

b) Why do plants need this gas?

The leaves do not have stomata on their top surface.
This ensures that they do not lose too much water.

The top of the leaf faces directly towards the Sun.
This means that it gets a lot warmer than the bottom.
If there were stomata on the top surface,
too much water would evaporate.

In most plants there is a layer of wax on the top.
This wax forms a waterproof layer.
This is not to stop water getting in, but to stop it evaporating.

If a plant loses water faster than it can be replaced, it **wilts**.

c) How does a plant replace water lost by transpiration?

This plant has lost too much water.

Remind yourself!

1 Copy and complete:

Most …… takes place in the leaves. Leaves are
broad and …… and directly face the …… Water
is lost from leaves by …… We call this process
…… Water leaves through the tiny holes called
……

2 Explain how carbon dioxide enters the leaves
and their cells.

3 Why is the wax layer on the top of leaves
important?

4 Explain why plants in hot/dry countries have a
thicker wax layer.

Have you ever done the washing at home?
Probably not!
But no doubt you've seen washing
drying on the line.

> **a)** What are the best conditions for drying washing?

What has all this domestic talk got to do with transpiration?

Clothes dry when the water on them **evaporates**.
Water is lost by evaporation from leaves.
So, a good drying day
is also a good transpiration day.

> Plants will lose more water on **dry**, **windy** and **warm** days.
> In these conditions the **rate** of transpiration is quicker.

In the right conditions a big tree could lose
nearly 200 litres of water in a single day.
This is the same volume as 600 cans of cola!

More transpiration occurs during the day than at night.
This is because light causes the stomata to open.
If the stomata are closed, then no water can be lost.

> **b)** Why do stomata close in very dry conditions?

This tree will lose many litres of water a day by transpiration.

Why is water so important to plants?

We know that water is needed for photosynthesis.
But it is also needed for support.

Plants that are short of water will **wilt** (go droopy).
When plant cells are full of water they are very firm.
This is because the water pushes against
the strong cell wall.

If all the plant's cells are firm, then the plant
will be well supported and stand upright.
Cells full of water are known as **turgid** cells.

This support is especially important for young plants.

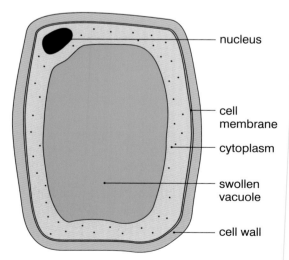

A turgid cell is full of water.

Why is transpiration so important?

Most photosynthesis goes on in the leaves.
Plants need water for this process and this
water comes from the soil.

> **c)** Why can't plants get their water
> through the leaves when it rains?

When water evaporates from the leaves,
it creates suction (a bit like a straw).
This draws water up the xylem tubes
to the cells of the leaves.
As water evaporates, more is drawn into the
roots.

> **d)** How does water enter the roots?

This continuous flow of water is called
the **transpiration stream**.
Without it water would not get to the
leaves where it is needed.

We know that transpiration is a very powerful force.
It is strong enough to pull water to heights
of over 100 m in Canadian Redwood trees.

Transpiration also has another useful job.
It helps to stop plants overheating.
When plants transpire it is just like us sweating.
When we sweat we lose water by evaporation.
Evaporation transfers heat and cools us down.

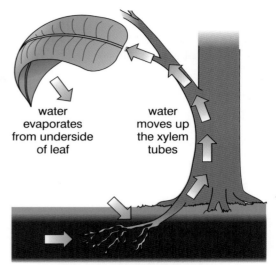

water evaporates from underside of leaf

water moves up the xylem tubes

water from the soil enters the roots

The transpiration stream.

Water is pulled right to the top.

Remind yourself!

1 Copy and complete:

The best conditions for transpiration are ......,
...... and ...... Light is also required, as the ......
close in the dark. Cells full of water are called
...... These cells help to ...... the plant. Plants
that are short of water ......

2 Explain why the transpiration stream is important
for photosynthesis.

3 Explain why trees do not need to depend on
water in their cells for support.

4 Draw a mind map to show what you know about
plant transport.

Summary

Plants produce food by **photosynthesis**.
Photosynthesis is summarised like this:
Carbon dioxide + water + (light energy) → glucose + oxygen
Chlorophyll is also needed to absorb the light energy.

The **rate** of photosynthesis can be limited by:

- shortage of light
- shortage of carbon dioxide
- low temperature.

The glucose that is made can be:

- used in respiration
- stored as starch
- used to make other useful chemicals, e.g. protein and cellulose.

Plants also need mineral salts like nitrates.
They absorb these from the soil through their roots.
Roots have tiny root hairs to give them a bigger surface area.

Leaves are well suited to photosynthesis.
They take in carbon dioxide through tiny holes called stomata.
These **stomata** can be opened and closed by **guard cells**.
They also let water vapour out.
This loss of water vapour is called **transpiration** and it is more rapid in windy, dry and warm conditions.

Plants have a circulatory system made up of:

- **xylem** vessels that carry water to the leaves
- **phloem** vessels that carry dissolved food from the leaves.

Water is important for support. Plants lacking in water wilt.
Water enters plant roots by **osmosis**.
Osmosis is the diffusion of water across cell membranes.

Questions

1 Copy and complete:

Unlike animals ...... make their own ......
Photosynthesis uses ...... energy to make ......
This process uses ...... and .... gas.
Plants also need a green substance called ......
to absorb ......

2 Plant cells take in water by osmosis.

i) What stops them from bursting open when they are full? (Hint: Look back at page 8.)

ii) Now explain why animal cells **do** burst open if they are put in water.

3 Greenhouses can be used to grow plants throughout the year.

i) How can growers provide extra carbon dioxide and heat at the same time?

ii) Explain why leaving the lights on all day will not necessarily make the plants grow better. Use the idea of limiting factors in your answer.

iii) Why must the temperature inside a greenhouse not get too hot?

4 a) Name the parts of the leaf labelled A to F.

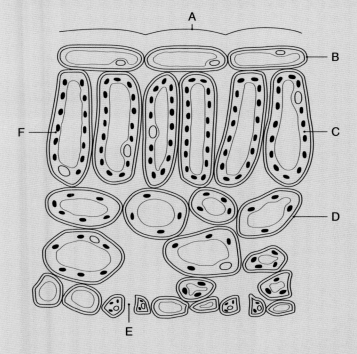

b) Why is the top of the leaf covered with a waxy layer?

c) Why does most photosynthesis take place in the palisade cells?

d) Why are there stomata only on the bottom of the leaf?

e) What do the xylem vessels bring **to** the leaf?

5 Early in spring the woodland floor is covered in flowers like bluebells.

Explain why these flowers need to appear **before** the trees develop their leaves. (Think about how plants need light.)

6 This apparatus can be used to collect oxygen gas from a water plant.

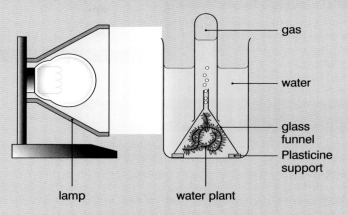

To alter the amount of light a lamp was placed at different distances from the plant.

The number of bubbles given off per minute was then recorded:

Distance from lamp (cm)	Bubbles per minute
100	5
80	10
60	15
40	25
20	30

i) Put these results on a line graph.

ii) Explain any pattern you see in the results.

iii) To keep this experiment a fair test what things must be kept the same each time the lamp is moved?

7 Explain the following observations:

i) Many plants store a lot of food in the autumn.

ii) Very little transpiration happens on winter days.

iii) Greenfly can often be seen feeding from the phloem tubes of a plant.

iv) Houseplants wilt very quickly when left on sunny windowsills.

CHAPTER 7

The nervous system

▶▶▶ 7a Detecting our world

In a number of places in Britain there are huge
radar stations like the one in the photograph opposite.

Many of them were built in the 1960s.
Their job was to detect attack by missiles launched
from Russia.
Today there is much less of a threat from Russia,
but these stations are still used to monitor events
all over the world. They do this by making use
not only of radar but also of satellite technology.

Animals need to detect what is going on around them.
To do this they have special cells called **receptors**.
These cells are designed to detect **stimuli**.

Stimuli are changes in our **environment** (surroundings).

a) What sort of changes do you detect during a typical day?

Receptor cells are not found just anywhere in the body.
They tend to be grouped together into organs.
We call these **sense organs**.

b) How many sense organs can you name?

This dog is detecting its environment.

Eyes

In your eyes you have **light-sensitive** cells.
These cells are sensitive to the brightness of light.
This allows you to tell when it is getting dark.
They also detect different colours.
These receptors send all this information
to your brain.
The brain turns all the information into the picture
you see.

Ears

Ears contain two types of receptor.
Each has a different job.

- Some of the cells are sensitive to sound.
They send information to the brain.
The brain then processes the information
into the sounds that you hear.
- Other cells deep inside your ear are sensitive
to changes in position.
They (with your brain's help) allow you to
keep your balance.

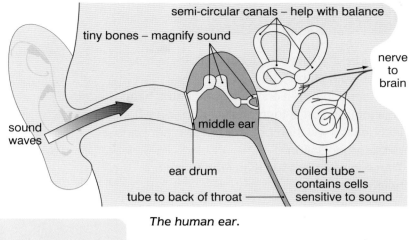

The human ear.

Labels: semi-circular canals – help with balance; tiny bones – magnify sound; nerve to brain; sound waves; middle ear; ear drum; tube to back of throat; coiled tube – contains cells sensitive to sound

c) Why do you sometimes feel dizzy if you
have a heavy cold?

Tongue and nose

These two organs are sensitive to chemicals.

- The tongue has hundreds of tiny receptors on
its surface. These are sensitive to the
chemicals in food and drink. Different parts of
the tongue are sensitive to different tastes, e.g.
sweet and sour.
- The nose does a similar job.
Its sensitive cells detect the chemicals in
gases. Both the nose and tongue enable us
to detect flavours in a meal.

*The nose and mouth are important for the
sense of taste.*

d) Why is your sense of taste affected
when you have a cold?

Skin

Finally, there are the receptors in your skin.
These are sensitive to a number of things.
They allow you to feel:

- pain when you injure yourself
- pressure – this helps when you pick things up
- temperature – so you know when something is
hot
- touch – so you can tell the difference between rough and smooth.

*Blind people read using their sense of
touch.*

Remind yourself!

1 Copy and complete:

To detect their surroundings animals have ......
...... These cells make up the .... Any
change in the ...... is known as a ......

2 Explain why, if you could not feel pressure, it
would be nearly impossible to pick up an egg.

3 Which part of the body coordinates all your
senses?

▶▶▶ 7b Coordination

The radar stations mentioned on the previous page
do not work alone.
They send information back to a central headquarters.
This information is then analysed to see what it means.
Instructions will then be sent out to other parts
of the organisation.
These instructions will then lead to actions
being carried out.

Your nervous system works in a similar way.
The receptors send information to the
central nervous system.

> The **central nervous system** is made up
> of the **brain** and **spinal cord**.

The brain analyses this information.
It then sends out instructions to **muscles** and **glands**.
These are the parts of the body that carry out **responses**.

One of the jobs of your brain and spinal cord is to
coordinate all of your body's activities.
This means it makes sure that all the different parts
work together in an organised way.

Imagine you are a footballer.
A ball is coming over from a corner.
To head a goal you need to coordinate your actions.
You need to time your jump just right to meet the ball.
Think about all the muscles that must be coordinated
to do this!

Even when we are asleep, coordination is important.
Swallowing, breathing and our heart beat must all
be coordinated to keep us alive.

a) Why do athletes in a relay race need good
hand–eye coordination?
b) What other sports need good
hand–eye coordination?

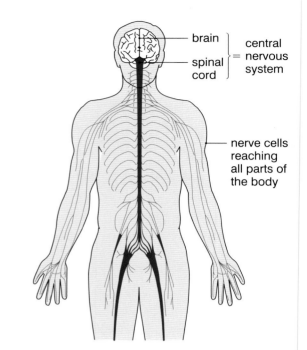

The human nervous system.

No coordination there!

Nerves

In Chapter 1 we saw that nerve cells are a
special kind of cell.
These cells have most of the features of typical
animal cells.
They have a cell membrane, cytoplasm and a nucleus.
The difference is in their shape.
The cytoplasm of nerve cells is drawn out
into a long, thin thread.
This thread is often called the **nerve fibre**.

A nerve cell.

> c) Why do nerve cells need to be long and thin?

There are two types of nerve cell (**neurone**).

> **Sensory** neurones carry information *from* the receptors
> **to** the central nervous system.

direction of nerve impulse

cell body

nerve endings
inside of receptor

nerve fibre

nerve endings inside
the central nervous
system

A sensory nerve cell.

> **Motor** neurones carry information *from* the central nervous system
> **to** the muscles and glands.

cell body inside
the central
nervous system

direction of nerve impulse

nerve fibre

muscle or gland

A motor nerve cell.

In both cases the information is carried along neurones
in the form of electrical signals (**impulses**).

Remind yourself!

1 Copy and complete:

   The brain ...... all of the body's activities.
   Information passes along ...... nerve cells from
   the ...... to the ...... It then sends messages
   along ...... nerve cells to the ...... and ......
   These structures carry out ......

2 What part of the body protects the delicate nerve
   cells from damage?

3 What is meant by the central nervous system?

4 In what way are nerve cells like telephone wires?

It is often said that to be good at sport
you need good reflexes.
But what are reflexes?
They are different from ordinary responses
because they are **automatic**.
By automatic we mean that they just happen.
You do not have to think about them.

A reflex response.

a) Name some human responses that are automatic.

Reflex actions are usually very rapid and they are designed to prevent injury.

- Coughing is a good example of a reflex.
  It is designed to remove objects from your windpipe.
  When food goes down the 'wrong way',
  vigorous coughing often removes it.
- Blinking is something else you do not usually think about.
  Combined with the liquid that makes up tears,
  blinking keeps the eyeball clear of dust and dirt.

In some sports, very fast responses are important.
In football, goalkeepers often make
what are called reflex saves, e.g. when they block a shot
from close range.
These are not really reflexes, it's just that the
goalkeeper has learned to respond very quickly.

Thank goodness for reflexes.

b) Why are fast responses important for sprinters?
c) What other sports need fast reactions?

Actions that we think about are often called
voluntary actions.
For example, picking up and reading this book.
Another word for a reflex action (one we do not think about)
is an **involuntary action**.

A point blank save.

The reflex arc

Reflexes involve most parts of the nervous system except the thinking bit of the brain.

d) How does this fact explain the speed of reflexes?

Look at the diagram below.
It shows the route taken by nerve impulses when you pick up a hot object.
This route is called a **reflex arc**.

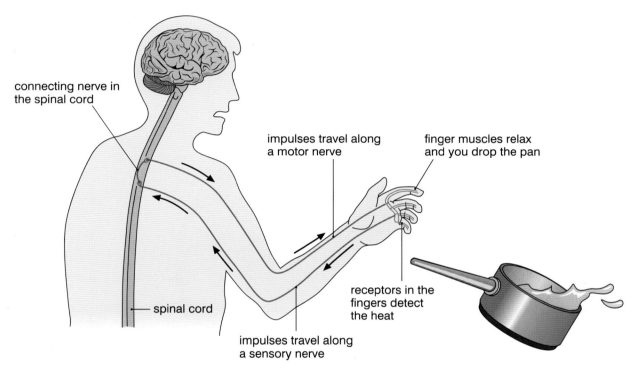

connecting nerve in the spinal cord

impulses travel along a motor nerve

finger muscles relax and you drop the pan

spinal cord

receptors in the fingers detect the heat

impulses travel along a sensory nerve

A reflex arc.

This is an example of a **withdrawal reflex**.
These reflexes protect the body from injury.

e) Can you think of another example of a withdrawal reflex?

Remind yourself!

1 Copy and complete:

Reflexes are …… responses. They are very …… and they protect us from …… The nerve impulses travel along a …… arc. Reflexes are very important in many …… Reflexes are also known as …… actions.

2 Copy and complete this table:

Stimulus	Reflex action
bright light	
dust in the eye	
food in the throat	
food in the windpipe	
touching a hot flame	

The eye is an example of a receptor.
It receives information about:
- the brightness of light
- the colour of light
- the shape of objects
- the movement of objects.

a) Make a list of all the things you could not do without the ability to see.

The list will be a very long one and should include reading this book!

Our eyes are very valuable so we must look after them.

The parts of the eye

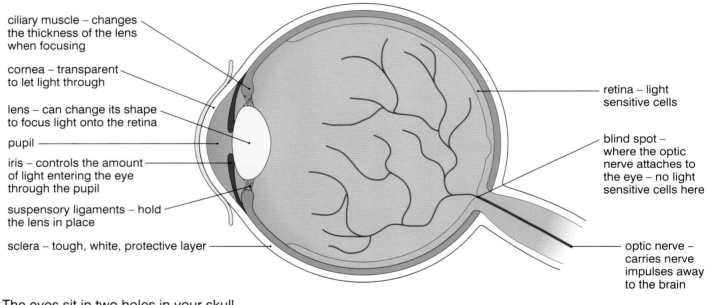

ciliary muscle – changes the thickness of the lens when focusing

cornea – transparent to let light through

lens – can change its shape to focus light onto the retina

pupil

iris – controls the amount of light entering the eye through the pupil

suspensory ligaments – hold the lens in place

sclera – tough, white, protective layer

retina – light sensitive cells

blind spot – where the optic nerve attaches to the eye – no light sensitive cells here

optic nerve – carries nerve impulses away to the brain

The eyes sit in two holes in your skull.
These holes are called **sockets**.
Your eyes are moved by three pairs of muscles.
These allow your eyes to swivel in their sockets.

How do we see?

- Light rays from an object you are looking at enter the eye.
- As they enter they are bent, by the **cornea** and also by the **lens**.
- The light rays are then **focused** on the **retina**. The retina contains thousands of **light-sensitive** cells.
- The light rays form a small, upside-down image on the surface of the retina.
- The light-sensitive cells now send this image (as nerve impulses) to the brain, along the **optic nerve**.
- The brain turns the image the right way up and allows you to see objects as they really are.

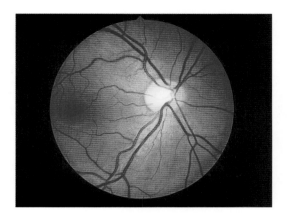

The light-sensitive retina.

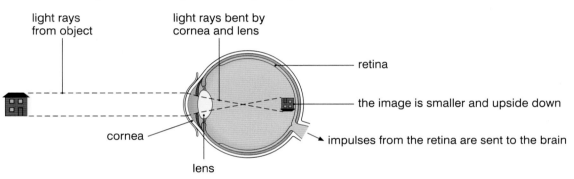

light rays from object

light rays bent by cornea and lens

retina

the image is smaller and upside down

cornea

impulses from the retina are sent to the brain

lens

How we see.

Two eyes are better than one!

Have you ever wondered why we have two eyes? With only one eye it is very difficult to judge distances. Try this exercise:

- Close one eye.
- Hold one finger out in front of you.
- Now try to touch this finger with a finger from the other hand.
- Repeat the exercise with both eyes open.

Which was easier, with one eye or two?

Judging distance is vital for safe driving.

Remind yourself!

1 Copy and complete:

Light enters the eye through the ...... It is focused on the ......, by the ...... and the ...... The retina has lots of light ...... cells.

The image on the retina is sent to the ......, along the ...... nerve.

2 Which part of the eye:

i) controls the size of the pupil

ii) holds the lens in place

iii) contains light sensitive cells

iv) has a transparent area?

3 Why are two eyes better than one?

Summary

The **nervous system** allows us to **react** to our **surroundings**.

We detect changes in our surroundings with cells called **receptors**.
These changes are called **stimuli**, and our reactions are called **responses**.

The brain **coordinates** all of our responses.

Nerve cells (**neurones**) are a special type of cell.
Sensory neurones carry **nerve impulses** to the brain.
Motor neurones carry impulses to muscles and glands.

Some responses are automatic and very rapid.
These are called **reflexes** and they protect us from injury.

The **eye** is an example of a **receptor**.
It is designed to focus light and form an image.
This image is transmitted to the brain along the optic nerve.
The brain interprets the image and allows us to see what we are looking at.

Questions

1 Copy and complete:

Our nervous system detects changes in our
...... These changes are called ...... The ......
receives information about these stimuli. It
then ...... the body's response to them. The
brain and the ...... cord make up the ......
nervous system. Information is passed around
the nervous system as ...... impulses. These
travel along ...... and ...... nerve cells.

2 Write down two examples of stimuli and the
responses that result from them.

3 How are nerve cells adapted to do their job?

4 What is the difference between a voluntary
action and a reflex action?

5 Where do we really 'hear' sounds and 'see'
objects?

6 When dogs smell their favourite food, what is
their (rather messy) reflex response?

7 Some animals (like rabbits) have eyes on the
sides of their head rather than towards the front.
What is the advantage of this?

8 This diagram shows the human eye:

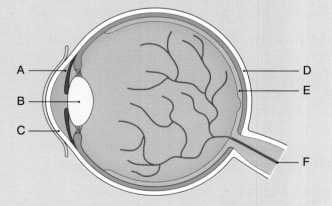

i)   Name the parts labelled A–F.

ii) Which of these parts focuses light?

iii) Which part carries information to the brain?

CHAPTER 8

Keeping control

Have you ever walked into an air-conditioned
building on a hot day?
The room seems cold at first.
That's because its temperature is being **controlled**
and compared to outdoors it is a lot cooler.

People are very good at controlling the conditions
they live and work in.
Even if your school isn't air conditioned
it will have central heating.
A central-heating system is operated
by a **thermostat**.

You can set a thermostat to any temperature you like.
If the temperature in the room falls, then it 'clicks'
the heating on.
If the temperature gets too high, it 'clicks' it off.

The human body works in a similar way.
It has ways of keeping its internal conditions
under control.

a) What is the normal human body temperature?

Controlling our internal conditions is important
for keeping our body working properly.
Apart from temperature we also need to control:
- the water content of our body
- the **ion** content of our body
  (e.g. sodium and chloride ions in salt).

To keep the body working we also need
to get rid of waste.

The two main waste products are:
- carbon dioxide gas
- urea – made in the liver when it breaks down
  unwanted amino acids.

b) How does our body make carbon dioxide gas?
(Hint: look at page 26.)

*In modern buildings conditions are carefully
controlled.*

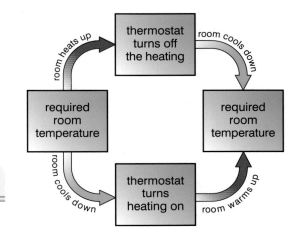

▶▶▶ 8a Drying out

Why is **water** so important?

Around 67% of the human body is water.

- All of the chemical reactions in cells take place in a watery solution. We can, in fact, live far longer without food than water.

- The amount of water in the body is controlled by the **kidneys**. The kidneys also remove poisonous substances made by cells.

- If the kidneys stop working, their jobs have to be carried out by a **dialysis** machine. If the kidneys do not start working again a **transplant** may be necessary. Replacing a diseased kidney with a healthy one is a very common and successful operation.

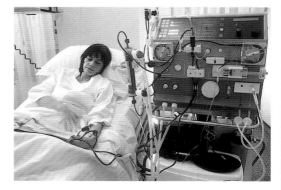

A dialysis machine does the job a healthy kidney does in most people.

a) Why do you think there is a waiting list for kidney transplant operations?

Where are the kidneys?

You have two kidneys in your lower back.

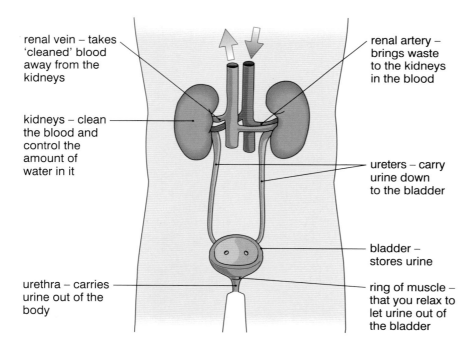

renal vein – takes 'cleaned' blood away from the kidneys

kidneys – clean the blood and control the amount of water in it

urethra – carries urine out of the body

renal artery – brings waste to the kidneys in the blood

ureters – carry urine down to the bladder

bladder – stores urine

ring of muscle – that you relax to let urine out of the bladder

Cleaning the blood

The poisonous substance that the kidneys remove is called **urea**.
Urea is made in the liver when **excess** amino acids are broken down.

> **b)** What are amino acids normally used for?
> (Hint: look at page 21.)

This chemical is dissolved in water to make **urine**.
Urine also contains excess ions (salts).
Urea and excess ions are removed from the blood by a process of **filtering**.

Deep inside the kidney there are thousands of tiny filter funnels. These work in a similar way to the filter funnels you use in chemistry. They only let small molecules through.

The kidney can remove varying amounts of water from the blood.
The brain contains cells that detect how much water is in the blood.

If we are **dehydrated** (short of water) it will 'instruct' the kidney to remove *less* water.
This is why on a hot day you produce less (and stronger) urine.

How else do we lose water?

We lose water from our lungs when we breathe out.
We also lose it from our skin when we sweat.
These losses are not easy to control.
This is why the kidneys have such an important role in controlling our water levels.

> **c)** Why do you urinate more often on cold winter days?

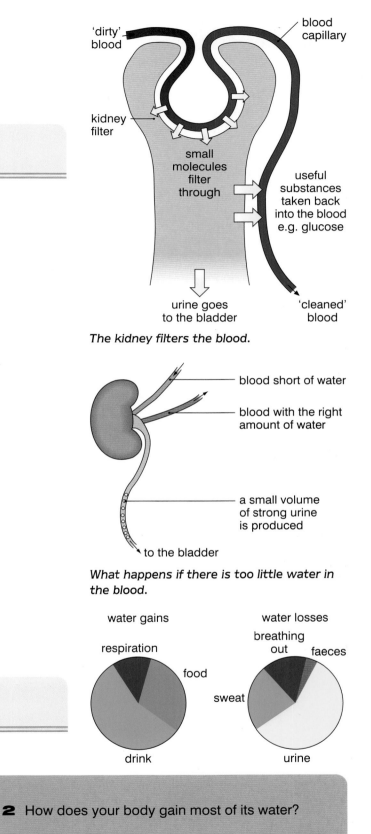

The kidney filters the blood.

What happens if there is too little water in the blood.

Remind yourself!

1 Copy and complete:

The ...... control the body's ...... level. They also remove ...... substances from the ...... They produce ...... which contains water, ... and ......

2 How does your body gain most of its water?

3 What are the two ways that someone with kidney disease can be treated?

Temperature control in the human body
is very important.
Every winter there are newspaper reports of
old people dying of **hypothermia**.

Our normal core body temperature is **37°C**.
If the temperature drops below 35°C, then
hypothermia develops. This condition causes
drowsiness, slows the heart and breathing rates
and may lead to death.
This condition does not just affect old people.
People of any age can be affected if they
are exposed to very cold conditions.

This person is in danger from hypothermia.

a) Why are old people particularly affected by hypothermia?

Humans are known as **warm-blooded animals**.
This means that we are usually able to keep
our body temperature steady at 37°C.

b) When are we often unable to keep
our temperature steady?

Our body makes heat because of all the chemical
reactions going on inside it.
This heat is carried around in our blood.
Our skin plays a big part in whether we keep
or lose this heat.
This diagram shows the structure of the skin:

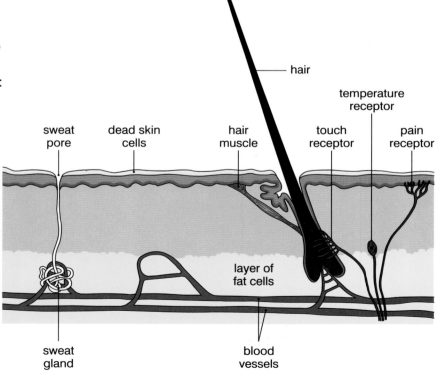

Cooling down

The brain detects when our body temperature
starts to rise.
If our temperature rises above 37°C, enzymes
in our cells will not work as well. Enzymes control
the chemical reactions in our body.

When you get too warm:
- You start to sweat more.
  When the sweat evaporates from your skin
  you will cool down.
- The blood vessels under your skin
  become wider.
  This lets more heat escape into the air.
- The hairs on your skin lie flat.
  This makes it easier for heat to escape.

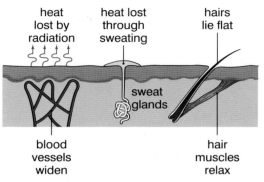

In hot weather.

c) Why do you go red when you are hot?

Keeping warm

When you get too cold:
- Sweating slows down.
- The blood vessels under your skin become
  narrower and let less heat out.
- The hairs on your skin stand upright.
  This traps warm air near to your skin.
- You start to shiver.
  Shivering occurs when your muscles start
  contracting and relaxing very quickly.
  This generates extra heat.

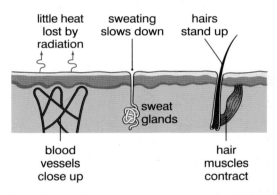

In cold weather.

Remind yourself!

1 Copy and complete:

The normal body temperature is …… If it rises
above this …… will not work properly. The ……
helps to control our temperature.

Changes in temperature are detected by the
……

2 Use the information on this page to explain why
you look pale if you are cold.

3 Apart from temperature control what are the
other functions of the skin? (Hint: look back at
the diagram on page 73.)

In Chapter 7 we saw how the nervous system
coordinates what the body does.
But not every process is coordinated by nerves.
The body has another system using chemicals.
These chemicals are called **hormones**.
Hormones play a big part in helping
the kidneys to control our water levels.

Hormones are made in parts of the body
called **glands** and are carried in the **bloodstream**.
Glands are parts of the body that release
chemicals into the blood.
Nerve impulses go straight to particular
parts of the body.
But because they travel around in the blood
hormones go everywhere.
The parts of the body they affect
are called the **target organs**.

This table summarises the differences between
the nervous and hormonal systems.

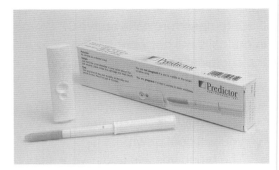

*Pregnancy testing kits work by detecting
hormones in urine.*

Nervous system	Hormonal system
carries electrical impulses	carries chemicals
impulses travel very quickly	chemicals travel more slowly
impulses affect specific organs	chemicals affect a number of organs
has short-term effects	has long-term effects

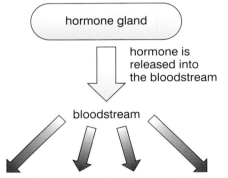

hormone gland

hormone is
released into
the bloodstream

bloodstream

hormone is transported to all parts of the body

Hormones can have long-lasting effects.
Therefore they are involved in controlling things
like growth and development.

It's their hormones!

Adolescence is the time of your life
when you undergo **puberty**.
This is when your body becomes sexually mature.
Adolescence is influenced by hormones and it
can be a very emotional time.
These hormones can make you feel irritable
and moody.
Don't worry – you grow out of it!

Where are hormones made?

This diagram shows the main hormones in the body and where they are made.

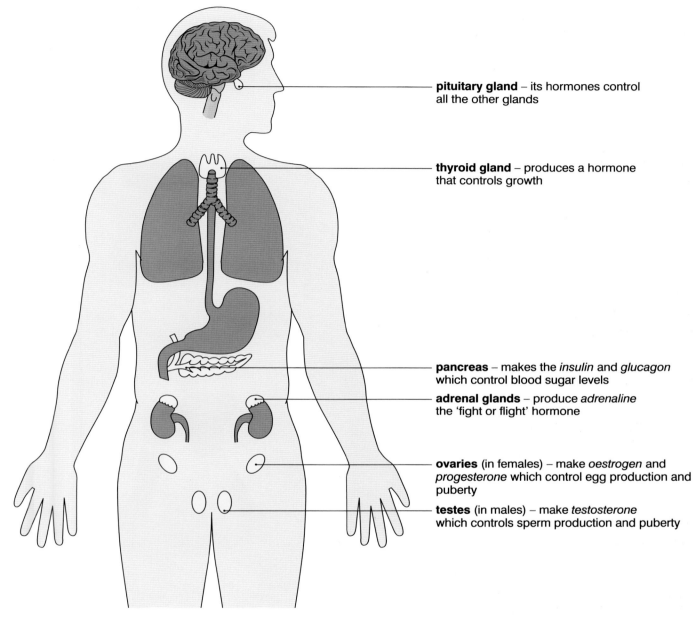

pituitary gland – its hormones control all the other glands

thyroid gland – produces a hormone that controls growth

pancreas – makes the *insulin* and *glucagon* which control blood sugar levels

adrenal glands – produce *adrenaline* the 'fight or flight' hormone

ovaries (in females) – make *oestrogen* and *progesterone* which control egg production and puberty

testes (in males) – make *testosterone* which controls sperm production and puberty

The hormone system.

Remind yourself!

1 Copy and complete:

Hormones are …… carried in the bloodstream. They help to …… the activities of the body. The parts of the body they affect are called the …… organs. Hormones are released into the blood from special ……

2 Why do hormones travel more slowly than nerve messages?

3 Which gland makes reproductive hormones in:

i) males

ii) females?

(Hint: look in Chapter 16.)

Do you know anyone who has diabetes?
They have a problem controlling the sugar (glucose)
in their blood.
The control of blood sugar is a good example
of hormonal control.

a) Why do our cells need sugar?
(Hint: look at page 26.)

It is important that our blood glucose level stays constant.
But some events make it change.
- After a meal it will rise as the food is
  digested and absorbed into the blood.
- During exercise it will fall as our cells
  use up sugar to release energy.

Diabetes

Blood glucose levels are monitored by the **pancreas**.

b) The pancreas is part of which body system?

Normally the pancreas releases hormones that control
the blood glucose levels.
- If the level is **too high** it releases **insulin**.
  Insulin causes the liver to take glucose out of
  the blood and store it in the liver. This makes
  the level of sugar in the blood fall.
- If the level is **too low**, it releases **glucagon**.
  Glucagon causes the release of glucose
  from the liver.
  This makes the blood sugar level rise again.

In most people this system works well.
People with **diabetes** have a faulty pancreas.
They cannot make the hormone insulin.
This means that after a meal there is nothing
to bring their glucose level back down.
Their glucose level could rise high enough
to cause death.

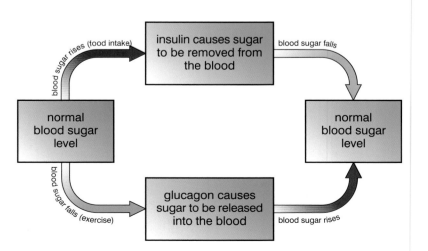

Treating the disease

Not being able to control blood glucose levels
can cause other problems.

Sometimes the level can fall too low.
This makes sufferers feel weak, irritable and confused.
Sometimes they will faint.
This condition is fairly easily treated with sugar.
If you know someone with diabetes, this is why
they carry glucose tablets or a drink around with them.

c) What is an incurable disease?

Diabetes is an incurable disease.
But it can be treated using injections of insulin.
These can be given by the patients themselves,
usually at meal times.
Diabetes sufferers also keep a careful watch on their diet.
They usually avoid high carbohydrate foods.

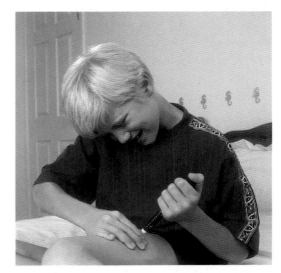

Insulin injections are used to treat diabetes.

Hormones and sport

Scientists have discovered other effects of hormones.

- **Anabolic steroids** are chemicals similar to **testosterone**.
  Testosterone controls male characteristics
  such as body shape and muscle.
  Some athletes use steroids to build up muscle.
  They also help to speed up recovery from injury.

- Using these drugs is illegal in sport.
  It gives athletes an unfair advantage over others
  who do not use them.

- There are also a number of side effects
  from using them.
  These include liver damage and heart disease.
  Also, in women athletes they can lead to some
  male characteristics.
  For example, the growth of lots of body hair.

A number of athletes have taken steroid drugs to improve performance.

Remind yourself!

1 Copy and complete:

Blood sugar is controlled by …… These
hormones are called …… and …… They are
made in the …… If the pancreas is faulty not
enough …… is made. This causes ……

2 Which part of the body is the **target** organ for
insulin?

3 Why do diabetics need to have insulin injections
at meal times?

We know that animals are sensitive
to lots of different stimuli.

a) Can you name three of these stimuli?

Plants are also sensitive but to just three stimuli:
* light
* moisture
* gravity.

Plants do not have a nervous system to
coordinate their responses.
But they do have a hormonal system.
In animals, hormones are carried in the blood.
In plants, hormones are carried by the same system
that carries dissolved food.

b) What is the name of the tubes that carry dissolved food
in plants? (Hint: look at page 60.)

Shoots seek out the light.

Have you ever wondered why shoots always grow upwards?
Also, what about roots, why do they always grow downwards?

Hormones are responsible for plants growing
the right way up.

Shoots

Shoots always grow upwards, *towards* the **light**.
In growing upwards they are also growing
against the force of **gravity**.

Roots

Roots always grow downwards, towards moisture
in the soil.
In growing downwards they are also growing
in the direction of the force of gravity.

The main type of growth hormone in a plant
is called **auxin**.
It is able to coordinate growth because
it affects shoots and roots in different ways.

These roots seek out scarce moisture.

The effect of auxin

This hormone **stimulates** (encourages) the cells
in shoots to grow more quickly.
However in roots it **inhibits** (slows down)
the growth of cells.

The response to light

When a shoot gets light from one side
it tends to grow in that direction.
The shaded side of the shoot has more auxin
than the light side.
The shaded side grows more quickly,
and so it bends towards the light.

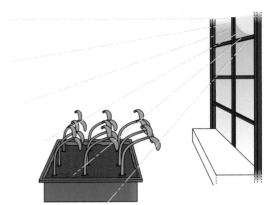

How shoots respond to light from one direction.

c) Why is it an advantage for a shoot to grow this way?

The response to gravity

Look at this drawing of a germinating seed
(one that is just beginning to grow).

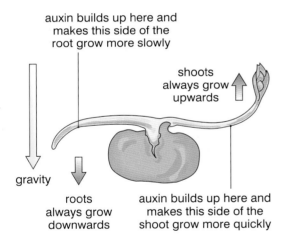

auxin builds up here and
makes this side of the
root grow more slowly

shoots
always grow
upwards

gravity

roots
always grow
downwards

auxin builds up here and
makes this side of the
shoot grow more quickly

Remind yourself!

1 Copy and complete:

Plants respond to ......, ... and ...... Shoots
grow ...... and roots grow ...... Their growth is
controlled by a ...... This hormone ...... the
growth of shoots. But it ...... the growth of roots.
The name of the hormone is ......

2 Why is it important that shoots grow upwards?

3 Why is it an advantage for roots to grow
downwards?

4 How could you make sure that a plant on a
windowsill grows straight?

● Have you noticed that in big supermarkets
you can buy ripe and unripe fruit
(e.g. yellow and green bananas)?
The unripe fruit is for ripening at home.
The ripe fruit is ready to eat.
Bananas are shipped in from thousands
of miles away.
So how can producers give us this variety?

Unripe bananas are green.

Unripe fruit is kept in cold storage.
This delays its ripening.
It can then be treated with hormones.
These hormones cause the fruit to **ripen**
just when required for sale.

a) Why is unripe fruit less likely to be
damaged in transport?

Plant hormones can also be used as **weed
killers**.
Farmers use growth hormones that only
affect plants with broad leaves.
Most weeds have broad leaves,
whereas crops like wheat have narrow
ones.

A weedkiller has been at work here.

The hormones are sprayed onto the weeds.
They cause the weeds to grow very rapidly.
This makes thin, spindly weeds that soon die.

A third use for plant hormones
is the growing of **cuttings**.
Taking cuttings is an easy way
of growing new plants from old ones.

A piece of stem with one or two leaves
is cut from a plant.
This is then dipped into a rooting powder.
This contains a hormone that encourages
roots to grow.
The cutting is then planted into soil
or compost and left to grow.

b) What are the advantages of growing
new plants from cuttings?

MURPHY

50g℮ promotes healthy growth
of many types of cuttings

Contains lots of lovely stuff

Dipping a cutting in rooting powder.

Summary

The human body produces two waste substances, **carbon dioxide** and **urea**.
Carbon dioxide is produced during **respiration**.
Urea is made when **excess amino acids** are broken down.

The internal conditions that have to be controlled include:
water content, ion (salt) content and temperature.

The **kidney** controls the water and ion levels.
Water is also lost in sweating and breathing.
The **skin** plays a big part in temperature control,
through activities like sweating and shivering.

A number of processes in the body are controlled with the help of **hormones**.
These are chemicals released (secreted) by glands and carried in the blood.

Blood glucose is controlled by two hormones, **insulin** and **glucagon**.
A lack of insulin causes the blood sugar level to rise dangerously high.
This condition is called diabetes.

The growth of plants is controlled by hormones.
They control how plants respond to light, gravity and water.
They can also be used to ripen fruit, kill weeds and help cuttings to grow.

Questions

1 Copy and complete:

   To work properly the body must get rid of
   ...... One waste product is ...... This is
   produced by the ......, from excess ....
   The body ...... must also be kept at ......°C.
   This is to make sure that ...... work properly.
   Many processes are controlled by chemicals
   called......

   These travel in the ...... and affect ...... parts
   of the body. Insulin is the hormone that
   controls blood ......

   Plants have hormones that control how
   they ......

2 Why do athletes have random blood tests?

3 Why is it dangerous for sportsmen and
   women to use hormones?

4 Find out which hormone helps us
   to fight or run away?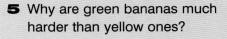

5 Why are green bananas much
   harder than yellow ones?

6 Why don't weed killers kill the crops as well
   as the weeds?

7 Why do diabetics carry glucose tablets with
   them?

8 Draw a mind map to summarise the
   information on plant hormones.

9 Why is it an advantage to have two kidneys?

10 Kidney failure can be treated by transplant
   surgery.

   a) What other organs are commonly
      transplanted?

   b) Why do these transplants not always
      work?

CHAPTER 9

DRUGS

▶▶▶ 9a What are drugs?

Drugs are often in the news,
unfortunately for tragic reasons.
One drug that has become very common
in recent years is ecstasy.
Ecstasy is a tablet often taken at parties
and nightclubs.
Users say that it makes them feel happy
and gives them the energy to dance all night.
Unfortunately a number of young people
have died after using ecstasy.

Many clubbers use ecstasy.

a) What other drugs have you heard of?

Making a list like this doesn't really tell us
what a drug is.

The word drug means:

> A chemical that affects the way the nervous system works.

They affect how you think, feel and behave.
Some speed up the nervous system.
These are called **stimulants**, and include
cocaine and caffeine.
Others slow the nervous system down.
These are called **depressants**, and examples are
heroin and barbiturates.

Coffee contains a stimulant, people drink it to keep alert.

b) Which type of drug do you think ecstasy is?

Some of these drugs have medical uses.
For example, heroin is used for pain relief in terminally ill patients.
But possessing them and using them for pleasure
is illegal in most cases.
Their use is dangerous because many of them
are **addictive**. This means that the body craves
more and more of them. Eventually a drug user
may be taking so much of a drug that its effects
can result in death.

This person is risking being infected by a virus.

Medicines

What is the difference between a medicine and a drug?

The word medicine covers a large group of chemicals.

Often they are called useful drugs.

Many of them work on the nervous system

For example:

- pain killers, such as aspirin
- anaesthetics used in operations to block pain
- sedatives, such as sleeping pills.

Some of these you can buy at a chemist, others you can only get from a doctor.

c) Which of the drugs above can you buy without a prescription?

A selection of useful drugs.

Not all medicines work on the nervous system.
Some are designed to kill bacteria.
If you develop a chest infection a doctor will often give you one of these medicines.
They are called **antibiotics**.

The most well known antibiotic is **Penicillin**.
Medical drugs like paracetamol can be dangerous if you take more than the prescribed dose.

d) What does the phrase 'prescribed dose' mean?

There are a number of everyday substances that are drugs.
Two of the most common are alcohol and nicotine.
Although legal above a certain age, they can also be very dangerous.

NATIONAL HEALTH SERVICE
G.P. PRESCRIPTION Serial Code: GHR65E

PLEASE PREPARE FOR: Mr. I.R. Jones

1 Course of 50 x 50 mg Penicillin

Doctor's Signature: J Smith

Remind yourself!

1 Copy and complete:

Drugs affect the ...... system. They can speed it up or ...... it down. Medicines are ...... drugs. Many drugs are ...... and users find it very difficult to ...... them up.

2 What is the difference between a stimulant and a depressant?

3 What is meant by the word addiction?

Tobacco is one of the few examples of a **socially acceptable drug**. This means a drug that is not illegal and can be used in public.

This baby is in danger of passive smoking.

a) What is the other main socially acceptable drug?

However, smoking is being banned in more and more public places. This is because of the health risks linked to smoking. Not just to smokers but to other people around them.

b) What do we call non-smokers who breathe in other people's smoke?

In Britain, thousands of people a year die from smoking related diseases. So just why is tobacco bad for you?

Tobacco smoke contains lots of harmful chemicals:

Nicotine

Nicotine is an addictive drug. This is why smokers find it very difficult to give it up. It also raises the heart rate and blood pressure. Over a long period of time this can damage the heart and blood vessels.

Tar

Tar is a thick and sticky substance that contains lots of poisonous chemicals. It collects in the lungs as the smoke cools down. Tar blocks the tiny tubes and air sacs and this makes breathing difficult. It is also responsible for causing lung cancer. This results in large groups of cells (tumours) that grow out of control in the lungs. Without early treatment this can often lead to death.

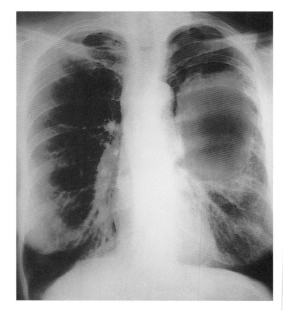

The red area on this X-ray is a lung tumour.

Carbon monoxide

Tobacco smoke contains a lot of carbon monoxide. This gas reduces the amount of oxygen the blood can carry. This explains why smokers are less able to play sport.

In pregnant women it also means that the fetus might get less oxygen than it needs.

This can lead to a baby being born under-weight.

Other lung diseases

Bronchitis

Tar in cigarette smoke causes cells in the breathing tubes to produce large amounts of sticky **mucus**. Normally mucus is cleared away by tiny hairs called **cilia**. Unfortunately, tar also damages these hairs. As a result the tubes fill up with mucus. The warm and sticky mucus is an ideal place for bacteria to breed. This build-up of bacteria causes **bronchitis**. People with bronchitis often cough a lot to try and clear the mucus.

c) Why is this often called a 'smoker's cough'?

Emphysema

Emphysema is also caused by chemicals in tobacco smoke.

The walls of the air sacs (alveoli) are destroyed. This means that there is a much smaller surface area for exchanging gases. People with this disease get breathless very easily. Eventually they may not even be able to get out of bed.

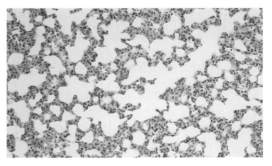

A healthy lung with lots of tiny air sacs.

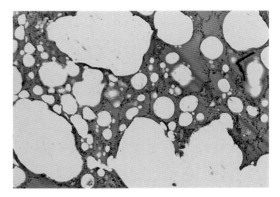

The air sacs in this diseased lung are enlarged and damaged.

Smoking while pregnant can harm the unborn baby.

Remind yourself!

1 Copy and complete:

Tobacco smoke contains many harmful ...... Nicotine is an ...... drug. This makes it ...... to give up smoking. Smoking is linked to many ...... People who do not smoke can still be affected, they are called ...... smokers.

2 Why is using nicotine gum safer than smoking?

3 Do some research to find out the latest figures for deaths caused by smoking.

Old habits die hard!

Alcohol is another example of a legal drug.
It is a much older habit than tobacco though.
Records show the Egyptians producing
alcohol around 4000 years ago!

Alcohol may be legal but it is still harmful.
It is another example of a **depressant** drug.

a) How do depressant drugs affect the body?

It is also an **addictive** drug.
People can become **dependant** on it
to get through life.
When this happens giving up alcohol
can be very difficult.
People who become dependant on
alcohol are called **alcoholics**.
When they try to give up they can suffer
unpleasant side effects.
These side effects are called
withdrawal symptoms.
Withdrawal symptoms are common
to most drugs in fact.

Alcohol can rule your life!

How does alcohol affect the body?

- Alcohol is taken into the body through
  the gut. It gets into the blood and is carried
  to the brain.
- A small amount of alcohol makes people
  relaxed and happy.
- Greater amounts affect their judgement
  and their reactions.
- With more alcohol people lose the ability to
  control their muscles. Their speech
  becomes slurred and they cannot walk
  properly.
- Large amounts of alcohol lead to unconsciousness.
  In severe cases a person may go into a **coma**.

. . . but it doesn't have to!

b) Why is it dangerous to drink and drive?

How much is too much?

Different amounts of alcohol will affect
people differently.
A lot depends on things like your body size
and whether you have eaten recently.
One way of comparing different drinks,
is to compare how many units of alcohol
they contain.
Look at the diagram opposite.

Doctors recommend that the safe limit
for men is 21 units per week.
The limit for women is 14 units per week.

½ pint beer (0.3 litre)   1 glass sherry   1 single whisky   1 glass wine   ½ pint cider (0.3 litre)

All of these drinks contain one unit of alcohol.

c) Why is it difficult to work out how many
units you are drinking?

How does alcohol damage the body?

Alcohol is a **poison**.
Drinking large amounts for many years
can damage the body.

- It causes the blood pressure to rise.
- This can cause heart disease.
- It also destroys brain cells.
- The worst damage can be to the liver.
  The liver is a very important organ.
  One of its jobs is to break down
  poisonous substances like alcohol.
  If the liver has too much alcohol to deal
  with, it can be damaged itself.
  Hepatitis and **cirrhosis** are two liver
  diseases common in heavy drinkers.

If the liver is damaged it is less able to
remove other poisons from the blood.
In severe cases liver damage can be fatal.

Alcohol can cause severe health problems.

Remind yourself!

1 Copy and complete:

Alcohol is a ...... drug. It ...... down the nervous
system. It is also a ...... that damages the body.
Alcohol is ...... and stopping ...... can be very
difficult. Alcohol can damage the heart and the
...... but the worst damage is to the ......

2 Why are spirits like whisky served in smaller
glasses than beer? Find out the percentages of
alcohol in some drinks.

3 Find out about an organisation called Alcoholics
Anonymous.

How do they help people with drink
problems?

Solvents are found in everyday products.
Things like glue, aerosol sprays and lighter fuel.
They give off fumes that can be breathed in.
The fumes quickly get into the blood and then
are carried to the brain.

Solvents have two main effects on the body:
- One effect is on behaviour.
  Sniffers can appear to be drunk.
  Their speech might be slurred and
  their coordination and concentration will be
  poor.
  Just as with being drunk these effects
  will make them more likely to have accidents.
- Solvents also cause damage to the body.
  In particular they damage the lungs,
  the liver and the brain.
  Solvents affect both the breathing rate
  and the heart rate.
  Sometimes inhaling a solvent can cause
  immediate death.
  This usually happens because the chemical
  causes heart failure.

A lot of solvent abusers tend to be
in the 13–16 age group.
It is illegal for shops to sell products like
glue to people under 18.
But because solvents are found in so many
products, it is a difficult problem to solve.

Using solvents has other dangers too.
Many solvents are sniffed from plastic bags.
This means there is a danger of suffocation.

Some abusers spray aerosols into the mouth.
This can freeze the air passages and again cause
suffocation.
Also many solvents are highly flammable,
so there is a big fire risk.

a) What are the main organs damaged
by solvents?

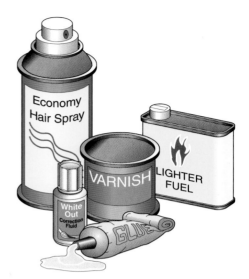

All of these products contain solvents.

Solvent abuse is very dangerous.

Summary

Drugs are chemicals that change the way our body works.
They affect the nervous system, which causes our behaviour to change.
They can also damage various parts of the body.
This damage can eventually lead to death.

Stimulants speed up the nervous system and **depressants** slow it down.

Tobacco smoke contains many poisonous chemicals.
Nicotine is **addictive** and it causes damage to the heart.
Tar also contains poisons and this contributes to lung diseases
like bronchitis, emphysema and cancer.
Tobacco smoke also contains **carbon monoxide** gas.
This prevents the blood from carrying enough oxygen to the cells.

Alcohol slows down the nervous system affecting our judgement and
coordination.
It also causes damage to the brain, heart and liver.
Long-term alcohol abuse can lead to death.

Solvents are found in many household products.
They are breathed in and can cause behavioural changes.
Solvents also cause damage to the lungs, liver and brain.

Many drugs are **addictive**.
This means that users become dependant on them.
Giving them up is very difficult and can lead to withdrawal symptoms.

Questions

1 Copy and complete:

Drugs affect the ...... Some drugs like ......
speed it up. others like ...... slow it down.
Drugs affect the body in two ways. They
change our ......, for example changing the
way we ...... and feel. Drugs also cause ......
to our body. The brain is often damaged and
so is the ......

Many drugs are ...... and giving them up is
hard. Some drugs like alcohol and tobacco are
......, but they are still harmful. Drug abuse is
the cause of many ......

2 Explain why non-smokers who work with smokers have an increased risk of getting lung diseases.

3 Why is alcohol called a 'socially acceptable' drug?

4 Find out the latest figures for deaths from solvent abuse.

Use this information to help design a poster warning teenagers of the dangers of using solvents.

5 Why are pregnant women advised to avoid smoking and drinking?

▶ Plant life

1 The diagram shows a plant leaf during photosynthesis.

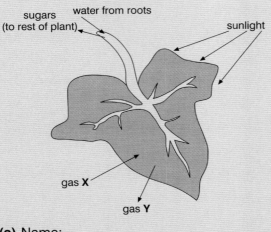

(a) Name:

   (i) gas **X**;

   (ii) gas **Y**. (2)

(b) Name the tissue which transports:

   (i) water into the leaves;

   (ii) sugars out of the leaves. (2)

(c) Why is sunlight necessary for photosynthesis? (1)

(AQA (NEAB) 1999)

2 Jason grows some plants.
He notices that they grow faster in the summer.

(a) Write down **two** reasons why plants grow faster in the summer. (2)

Jason looks at a root from one of the plants.
He uses a microscope.
Look at the diagram.
It shows some of the root cells he sees.

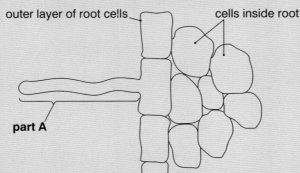

(b) (i) Write down the name of **part A**. (1)

   (ii) Describe **two** jobs of roots. (2)

(OCR 1999)

3 The rate of photosynthesis of a well-watered plant in a glasshouse was measured at different light intensities.
The atmosphere of the glasshouse had a high concentration of carbon dioxide.
The temperature was constant at 15°C.
The results are shown in the graph.

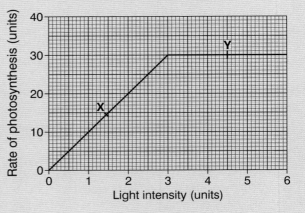

(a) What was the maximum rate of photosynthesis measured in this investigation? (1)

(b) What factor was limiting the rate of photosynthesis:

   (i) at point **X**;

   (ii) at point **Y**? (2)

(AQA 2001)

4 Photosynthesis is the process by which plants produce food.

(a) Copy and complete this word equation for photosynthesis:

$$\text{carbon dioxide} + \ldots\ldots \xrightarrow[\text{chlorophyll}]{\text{sunlight}} \ldots\ldots + \text{oxygen} \quad (2)$$

(b) Name **three** minerals essential for plant growth. (3)

(CGP Sample Question)

5 Jason measures the amount of water taken in by a plant.

He also measures the amount of water lost through the leaves.

His results are in the table.

Time	Morning			
	8 am	9 am	10 am	11 am
water taken in by the plant in cm³ per hour	10	12	12	13
water lost through the leaves in cm³ per hour	4	7	11	13
appearance of the plant	leaves flat and straight			

Time	Afternoon			
	12 noon	1 pm	2 pm	3 pm
water taken in by the plant in cm³ per hour	15	18	21	24
water lost through the leaves in cm³ per hour	17	22	28	32
appearance of the plant	leaves floppy (wilting)			

Looks at the results.

(a) (i) Explain why the leaves were flat and straight in the morning. (1)

(ii) Explain why the leaves were floppy (wilting) in the afternoon. (1)

(b) The leaves also make food.
Write down the name of this process. (1)

(OCR 1999)

6 A student grew two sets of cress seedlings.
Set **A** had light coming from all directions.
Set **B** had light coming from one side only.

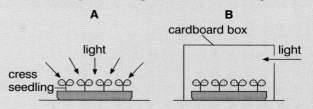

(a) Copy the drawings below and draw the appearance of the seedlings after two days in these conditions.

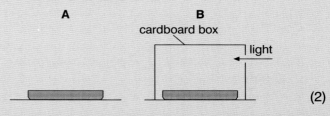

(2)

(b) Copy and complete the sentences about plant responses.
The response in **B** is the result of unequal distribution of …… in the seedlings.
This causes unequal …… . (2)

(AQA (NEAB) 1999)

7 Hormones control different features in plants.
Look at the list. It shows some features of plants.

A resistance to disease
B fruit ripening
C dark green leaf colour
D being poisonous
E producing flowers

Scientists know that **two** of the features on the list are controlled by hormones.

(a) Write down which two.
Choose the correct letters from the list. (2)

(b) Some plant hormones can make plants grow quicker.
Why would hormones like these be useful to a farmer? (1)

(OCR 1999)

Further questions on the Maintenance of life

▶ The nervous system

8 **(a)** Copy and complete the table to show which sense organ is linked with which sense.

Sense	Sense organ
touch	skin
balance	
	eye
taste	
smell	

(4)

(b) Sense organs collect information about our surroundings.
Why is it important to be able to collect the information? (1)

(OCR 1999)

9 **(a)** The diagram shows a reflex arc.

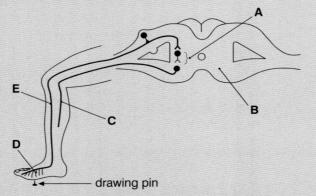

drawing pin

Copy and complete the table by matching the letters of the diagram to the part of the reflex arc. The first one has been done for you.

part of reflex arc	letter
relay neurone (nerve cell)	A
receptor	
sensory neurone	
spinal cord	
motor neurone	

(3)

(b) How do reflex actions help to protect our bodies from damage? (1)

(OCR 2001)

10 **(a)** Copy and complete the table about receptors. The first answer has been done for you.

Receptors in the	Sensitive to
Eyes	Light
Skin	
	Sound
Tongue	

(3)

(b) The diagram shows a section through the eye.

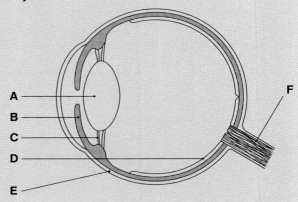

Which of the letters, **A** to **F**, on the diagram is:

(i)   the iris;

(ii) the sclera;

(iii) the retina? (3)

(c) Describe, in as much detail as you can, how information is transmitted from light receptors in the retina to the brain. (3)

(AQA (NEAB) 1999)

▶ Keeping control

11 The diagram shows some of the organs which keep the conditions inside the body constant.

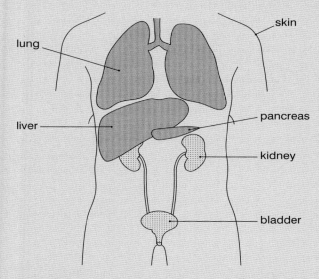

Use labels from the diagram to complete these sentences.

The ...... gets rid of the carbon dioxide we produce.

Urea is made in the ....

Urine is made in the ....

The ...... produces sweat.

Insulin is made in the .... (5)

(AQA (NEAB) 2000)

12 The table shows four ways in which water leaves the body, and the amounts lost on a cool day.

	Water loss (cm³)	
	Cool day	Hot day
breath	400	the same
skin	500	
urine	1500	
faeces	150	

(a) (i) Copy and complete the table to show whether on a hot day the amount of water lost would be

less   more   the same

The first answer has been done for you. (3)

(ii) Name the process by which we lose water from the skin. (1)

(b) On a cool day the body gained 2550 cm³ of water.
1500 cm³ came directly from drinking. Give **two** other ways in which the body may gain water. (2)

(AQA (NEAB) 1999)

13 The table shows the volume and the composition of the urine that five students, **A**, **B**, **C**, **D** and **E** produced during 24 hours. All the students drank the same amount of liquid in the 24 hours.

Student	Total volume of urine (cm³)	Amount of glucose in urine (g)	Amount of urea in urine (g)
A	1650	0	35
B	1830	8	22
C	1710	0	24
D	875	0	20
E	1680	0	16

(a) Which organ in the body produces:

(i) urine

(ii) urea. (2)

(b) Which student's urine contained the highest concentration of urea? (1)

(c) Which student had probably been the most active during the 24 hours?
Explain the reason for your answer. (2)

(d) (i) Which student was probably suffering from diabetes? (1)

(ii) Give **two** treatments for diabetes. (2)

(e) Which student had probably eaten most protein during the 24 hours?
Explain the reason for your answer. (2)

(AQA (NEAB) 2000)

▶ Drugs

14 Emphysema and tuberculosis are lung diseases. The bar charts show the link between cigarette smoking and these diseases.

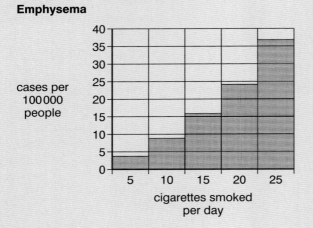

Emphysema

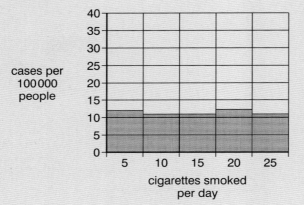

Tuberculosis

Look at the charts.

(a) How does cigarette smoking affect each of these two lung diseases? (2)

(b) Smoking can cause other diseases. Look at the list of diseases.

> **athlete's foot   cancer   diabetes**
> **heart disease   influenza (flu)**
> **sickle cell anaemia**

Write down **two** diseases which can be caused by smoking.
Choose your answers from the list. (2)

(OCR 1999)

15 (a) Sam has a chest infection. The doctor gave her antibiotics. How do antibiotics work to cure a chest infection? (1)

(b) Some drugs, when mis-used harm the body. Copy and complete the table by putting a tick [✓] under alcohol or solvents or both to show the effects of these drugs. One has been done for you.

Drug		Effect
Alcohol	Solvents	
✓	✓	possibility of addiction
		hallucinations
		damage to brain and liver cells
		damage to kidney cells

(2)

(c) It is dangerous to drink alcohol and then drive or operate machinery. Explain why. (2)

(AQA (SEG) 2000)

16 The table gives information about the amounts of alcohol consumed and the number of deaths from liver disease in some countries.

Country	Mean amount of alcohol consumed per person per year (litres)	Number of deaths from liver disease per 100 000 people per year
England	8	4
France	17	34
Germany	13	27
Iceland	4	1
Spain	15	22
Sweden	6	12

(a) Draw a bar chart to show this information.

(b) Alcohol causes harm to the liver. Explain how the information provides evidence of this. (1)

(c) The mean alcohol consumption is lower in Germany than in Spain. However, the death rate from liver disease is higher in Germany than in Spain. Suggest **one** reason for this.

(1)

(AQA (NEAB) 1998)

Section Three
Environment

In this section you will find out about how living organisms are adapted to their environment.
You will learn about how energy is transferred between living organisms and how useful chemicals are recycled.
You will also look at how humans affect the environment.

CHAPTER 10    HABITATS

▶▶▶ 10a The environment

You are probably reading this book in a classroom.
At the moment this room is your **environment**.
Soon your environment will change.
Your science lesson will end and your environment
might become the canteen or playground.

> The word 'environment' describes your surroundings.

The environment is made up of two parts, **living** and **non-living**.
The living part is all the plants and animals.

a) What living things are found in your classroom?

You probably mentioned your teacher and classmates.
But did you include the bacteria on the benches and
the plants on the windowsill?

The non-living part of the environment is all
the **physical** things.
These include things like temperature and
the availability of water and gases like oxygen
and carbon dioxide.

The world we live in is made up of lots of different **ecosystems**.

> An ecosystem is a group of living things
> and their particular physical environment.

A freshwater pond is one example of an ecosystem.

The Earth – a collection of ecosystems.

Look back at the drawing of the pond

b) Name two of the parts that make up
its physical environment.

The photograph opposite shows another ecosystem.
This is a woodland.

c) Name three animals you might find in a woodland.

Some of these animals will live in the trees.
Others will live on the woodland floor.

> The place where an animal or plant lives
> is called its **habitat**.

A habitat like a tree might be home to a number of animals.
There might be squirrels, owls, bark beetles and woodpeckers.

All the owls living in the wood make up a **population**.

> A population is made up of all the living organisms of one type
> living in the same place.

All the plants and animals living in the wood
make up its **community**.

> A community is all the different plant and animal populations
> living in the same place.

d) What is the difference between a population
and a community?

Remind yourself!

1 Copy and complete:

The word …… describes our surroundings. It is
made of two parts, living and …… An …… is a
group of …… things and their particular
environment. A …… pond is a good example of
an ecosystem. In a pond some animals live in
the …… at the bottom. This is their ……

2 Name one other land and one other water
ecosystem.

3 Apart from food, in what other way do animals
benefit from the plants in a woodland?

4 Apart from size, what is the biggest difference
between a freshwater and a marine ecosystem?

Despite the British hobby of moaning about the weather, we have a fairly comfortable climate. It is rarely too hot, too cold, too dry or too wet.

The conditions in other parts of the world are less kind.
Desert regions are very hot and dry, and the Arctic is very cold.
Despite these conditions animals and plants *do* live there.

Our weather is not usually this bad!

a) Name an animal that lives in the desert and one that lives in the Arctic.

These animals have special features that help them survive in these conditions.

> We call these special features **adaptations**.

A good example of a desert animal is the camel.
It has lots of adaptations to help it to survive in hot and dry conditions.

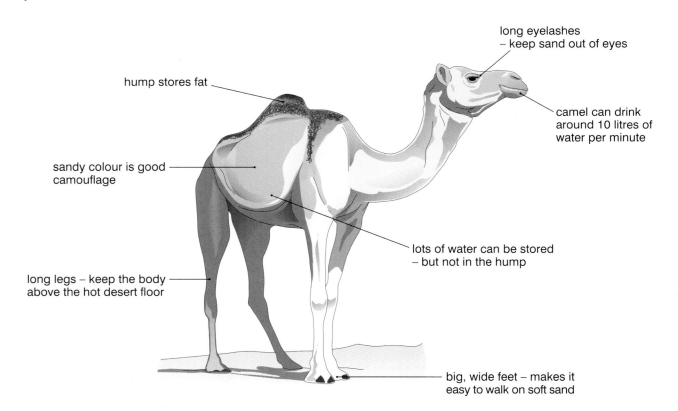

long eyelashes
– keep sand out of eyes

hump stores fat

camel can drink around 10 litres of water per minute

sandy colour is good camouflage

lots of water can be stored – but not in the hump

long legs – keep the body above the hot desert floor

big, wide feet – makes it easy to walk on soft sand

Surviving the big freeze

The polar bear is a good example of a creature adapted to live in very cold conditions.

white fur is good camouflage

thick fat layer under the fur for warmth and as a food store

thick fur which is warm and waterproof

small ears to prevent excess heat loss

strong legs for swimming and running

big, wide feet stop it sinking into the snow

big claws for killing prey

Body size and surface area

The surface area of an animal's skin is very important. It has a big effect on how much heat is lost.

The polar bear has a very bulky body (**a large volume**). But because it has a rounded shape its skin area is not as big as you might expect. This means that it does not lose too much heat.

The camel has a lot of loose skin. But the volume of its body is not that great. This large skin area helps it to lose heat easily.

Remind yourself!

1 Copy and complete:

Animals have …… to the conditions they live in. These adaptations are very important in …… environments. In a desert …… and a lack of water are big problems. In the Arctic the low …… is the main difficulty.

2 Why are polar bears white and camels brown?

3 Seals also live in the Arctic. Despite the cold they have no hair. So how do they keep warm?

4 Find out the highest temperatures and lowest temperatures that animals can survive in.

It is not only desert and Arctic animals that have special adaptations.
There are plenty of examples that are found in Britain.

The dormouse is an example of an animal that **hibernates** in the winter.

During hibernation an animal's temperature, breathing and heart rate fall to a low level.

The dormouse survives the cold winter using food reserves built up in the autumn.
It eats lots of food rich in carbohydrate.
These foods are then stored as fat.

a) Name another animal that hibernates.

Another way of avoiding the cold is to **migrate**.
Migration involves moving to a warmer continent during the winter.
Obviously this involves travelling long distances.
Therefore it is most common amongst birds.
For example, the swallow migrates from Britain to South Africa during our winter.

b) Why is South Africa a good place to go in our winter?

Migration is one way of surviving when the weather gets cold.

Another good example is **huddling**.

Emperor Penguins spend a lot of time on the freezing ice of Antarctica.
The penguin is a bird, but it cannot fly and therefore it cannot migrate.
To keep warm the penguins stand in large groups.

The birds on the outside gradually swap places with those on the inside.
In this way all the birds get a share of the warmth that is found in the middle of the **huddle**.

c) Which part of the Earth does Antarctica cover, the north pole or the south pole?

Plants have adaptations too!

Animals are not the only living things
that have to survive harsh conditions.

There are not many plants in a desert.
Those that do live there have to survive
in very dry (**arid**) conditions.

The best example of this type of plant is the **cactus**.

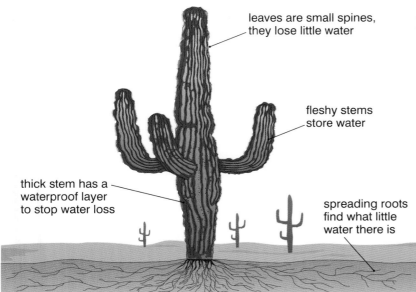

leaves are small spines,
they lose little water

fleshy stems
store water

thick stem has a
waterproof layer
to stop water loss

spreading roots
find what little
water there is

Another plant with special leaves is the **conifer** tree.
These trees are often found high up on mountains.
The leaves of a conifer are made up of small needles.
Just like the cactus these leaves are designed
to reduce water loss.
This is an advantage because it can be very windy
on mountainsides.
The windier the weather the more water will evaporate
from the leaves.

d) What is the proper name for evaporation
of water from leaves? (Hint: look in Chapter 6.)

Remind yourself!

1 Copy and complete:

Many animals have special ways of …… in cold
weather. Some ……, which involves a deep
sleep. Some birds ……to warmer countries.
Penguins ……together in a group to share their
body warmth.

2 Why do plants growing on sand dunes have very
long roots?

3 The flounder is a flat fish that can change its
colour to match its background.

What is the advantage of this?

Humans love competitions.
They compete to be the fastest,
the stongest, the richest and
the most beautiful!

In nature, animals compete with each other
for **resources** to stay alive.
These resources are often in short supply.

Everyone wants to win!

a) Write down a list of things that animals compete for?

How big is your list?
Hopefully it includes these three important resources,
food, **water** and **space**.

It is easy to understand why food and water are important.
But what about space?
Obviously, too many animals in one place will affect
the supply of food and water.
But animals also need space to mark out their **territory**.
Without a territory animals will not attract a mate
and will not be able to breed.

These stags are fighting for mates.

There are two types of competition:
● competition between animals of the same kind (species)
● competition between animals of different species.

Animals fighting for mates is a good example
of competition between animals of the *same* species.

b) Write down another example of this kind of competition.

The biggest and strongest animals in any habitat
will usually win the competition.
These animals are more likely to survive and go on to breed.
The weaker animals will not get the resources they need.
These animals are less likely to breed and more likely to die.

c) Read the last part of this page again.
Explain how competition will affect the
size of a population.

A tale of two squirrels

Animals of different species also compete
for the resources they need.
Obviously they do not compete for mates,
but food and water are still very important.

> If two different species need exactly the same
> resources they cannot usually live together.
> One of the species will lose the competition
> and be driven away.

The red squirrel.

In Britain today there are two kinds of squirrel.
The red squirrel is **native** to Britain.

d) What does 'native' mean?

The grey squirrel was brought to Britain from America.
At the time (around 1870) the red squirrel was common
throughout Britain.
It lived successfully in both **deciduous** and **coniferous**
woods.
A deciduous wood is one where the trees lose their
leaves in the autumn.
Within about thirty years the grey squirrel had spread
throughout the country.
Today the red squirrel is mainly found in coniferous woods,
although many can still be found on the Isle of Wight.

The larger grey squirrel.

e) Why has the red squirrel survived successfully on the Isle of Wight?

Red and grey squirrels eat acorns and hazel nuts.
Grey squirrels can digest this food better than red
squirrels.
This is why they have been more successful.
The red squirrels are smaller and nimbler than grey
squirrels.
They can climb pine trees and feed on pine cones.
This is why red squirrels are still found in pine forests.

*Coniferous woodland – the main home of
the red squirrel.*

Remind yourself!

1 Copy and complete:

Animals ...... with each other for scarce ......
The most important resources are space, ...
and ......

In the competition the biggest and ...... animals
will always win. The winners are much more
likely to ......

2 Why is it a bad idea to keep too many gerbils in a
cage?

3 Why do wild animals fight with each other?

4 Find out why grey squirrels were brought to
Britain?

Have you ever watched a bird table in winter?
As soon as food has been left on it
a variety of birds swoop down on the table.
This is because in winter, food is a very rare resource.

This is another good example of animal competition.
But it doesn't just happen in winter.
Even in summer birds will compete with each other
for food.

a) What sort of food do birds eat?

Two animals that eat **similar** food *can* sometimes
live in the same place.
The blackbird and song thrush are common
in British gardens.
They both eat a variety of insects plus some berries.
There is one important difference in their diet.
Song thrushes eat snails and blackbirds do not.

b) Why are snails difficult to eat?

A blackbird.   *A songthrush.*

Song thrushes have developed a way of breaking open
the snail's hard shell.
- They pick up the snail in their beak and smash the shell open on a stone.
- They can then eat the soft parts of the snail.
- Blackbirds have not developed this skill.

The stones used to break open the snail's shell
are often called 'thrushes' anvils'.
You can recognise them by the collection of broken shells
scattered around them.

There is another slight difference in the diet
of these two birds.
Blackbirds tend to eat more berries than song thrushes.

These two differences explain why both birds
can survive in the same habitat.

c) What is the difference between the diet
of a blackbird and a song thrush?

Plants compete too!

It is not only animals that compete with each other.
Have you ever seen a wheat field?
The golden yellow of the wheat is often mixed
with bright red poppies.
The poppies will be competing with the wheat plants.

Plant competition.

d) What resources do plants compete for?

Your list is probably similar to the list for animals.
For plants space is important so that they can get
enough light on their leaves.

e) Why is light important for plants?

> Plants that cannot get enough light, water or nutrients
> will not grow well.

The best competitor in the plant kingdom is the **weed**.
A weed is simply a plant growing where it is not wanted.
Look at this diagram to see why weeds
are such good competitors:

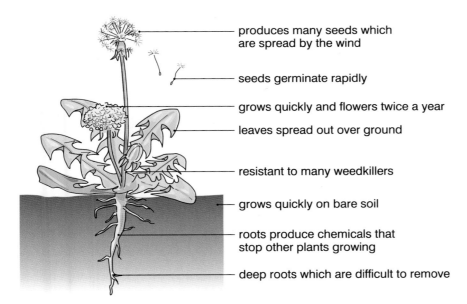

produces many seeds which
are spread by the wind

seeds germinate rapidly

grows quickly and flowers twice a year

leaves spread out over ground

resistant to many weedkillers

grows quickly on bare soil

roots produce chemicals that
stop other plants growing

deep roots which are difficult to remove

Remind yourself!

1 Copy and complete:

Plants compete for light, …… and …… Light is
needed for …… Plants use this process to make
…… …… are the best plant competitors.

2 Write down two reasons for removing weeds
from your garden.

3 Use the drawing above to explain why weeds are
difficult to remove from the soil.

Competition is one example of how nature stops populations from getting too big.

a) What do we mean by a population?
(Hint: look back at section 10a.)

There are lots of other reasons that explain why populations are kept in **check**:

- **Disease**: when populations become very crowded diseases can spread very quickly. Many individuals can be killed in this way.

- **Lack of shelter**: when animals do not have proper shelter they are exposed to danger.

- **Lack of space**: if animals do not have the space to mark out a territory, they may not breed.

- **Climate**: floods, storms and extreme temperatures can lead to the death of many plants and animals.

- **Loss of habitat**: forest fires and the cutting down of hedgerows destroy the habitat of many plants and animals.

- **Poisonous waste**: all living things give out waste chemicals. If these chemicals build up they can limit the size of a population.

- **Predation**: many animals catch and kill other animals for food, e.g. lions keep the numbers of antelope down.

- **Grazing**: grazing animals like sheep stop trees and shrubs from taking over grassland.

b) Which of these checks will have the same effect no matter how big the population?

How do populations grow?

If the conditions are right populations will grow very quickly.

- Around two hundred years ago Europeans first began to live in Australia.
  They took with them European rabbits.
- At first there was plenty of space, plenty of food and few predators.
- So the rabbit population grew very quickly.
- In just ten years twenty or thirty rabbits had grown into over twenty million!
- These rabbits destroyed many plants.
- They also competed with native animals for food.
- In the 1950s humans deliberately introduced a disease called **myxomatosis**.
- This very nearly wiped the rabbits out.

Healthy rabbits eat lots of plant life.

c) Why did humans want to kill the rabbits?

Without human interference most of the time populations just level off.
This is due to the factors that check populations mentioned on the previous page.

A growth curve is a graph that shows how the number of individuals changes over time.

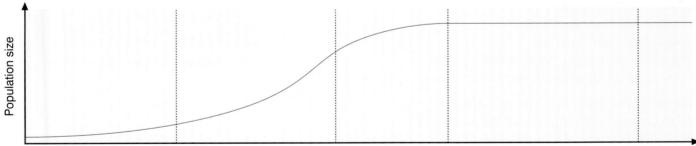

Population size / Time

few individuals so the population grows slowly

population grows very quickly

population growth slows down due to limited resources like food

number of births equals number of deaths
population size stays steady

Remind yourself!

1 Copy and complete:

In the right conditions populations …… very quickly. A number of natural …… stop populations becoming too …… These checks eventually mean that the size of a population …… off.

2 Why were rabbits able to spread so quickly in Australia?

3 Find out how the Chinese government is trying to limit the size of its country's population.

When did you last visit a fish and chip shop?
Most people love a good fish supper.
But do you realise that when you are eating
your cod and chips you are being a predator?

> A **predator** is an animal that catches and kills
> its own food.

Ok, so you are not actually doing the catching
and killing, but fishermen are doing it for you.
Humans are very good predators.
So good in fact that many species of fish
have nearly been wiped out.
The fish we catch are called the prey.

> **Prey** are the animals that predators catch and kill.

> **a)** Name some animals that are predators,
> and for each one suggest what its prey is.

Good predators are usually big, strong, fast
and aggressive.
They also have a number of tactics
that help them to be successful:
● they often hunt in packs
● they pick out young, weak or elderly prey
● they prey on more than one type of animal.

The cheetah – an excellent predator.

Successful prey animals (those that survive)
need to be well adapted to avoid the predators.
Some of the features of a good survivor are:
● good camouflage to avoid being seen
● good all-around vision to spot the predators
● the ability to run faster than the predator
● a horrible taste; this puts predators off
● staying in a large group for protection.

*Moving around in groups helps to avoid
predators.*

Predator–prey cycles

Predators do not only kill old or weak animals. Good predators will catch plenty of healthy animals too. This means that eventually the number of prey animals will fall to a low level.
We must remember that availability of food will also affect the size of the prey population.

With many predators and prey their numbers go up and down every few years or so.

We can show this **predator–prey cycle** on a graph like the one opposite:

1   There are lots of prey, so the predators breed and increase in number.
2   There are now lots of predators. So lots of prey get eaten and their numbers fall.
3   There is now less food for the predators. So their numbers start to fall.
4   With fewer predators around the prey start to breed. Their numbers increase and the cycle starts all over again.

A good example of this cycle is the lynx and the snowshoe hare.
These animals are found in Canada, and the lynx preys on the hare.

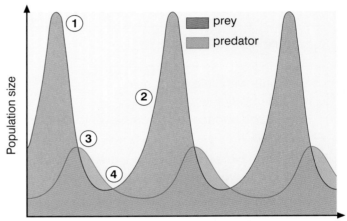

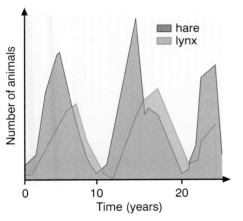

Summary

The surroundings of a plant or animal are its **environment**.
The environment is made up of **living** and **non-living** parts.
The non-living part (**physical**) includes things like air, water and soil.
The living part includes the **community** of the other plants and animals.
The **community is made up of lots of populations of animals and plants**.
A habitat is the place where an animal or plant lives.

Living things are **adapted** to survive in the conditions in which they live.
Animals are adapted to survive in extreme cold by having:

- thick coats
- a lot of body fat
- a small surface area compared to their volume.

In extreme heat animals have:

- coats with little fur or hair
- little body fat
- a large surface area compared to their volume.

Plants also have adaptations to extreme dryness,
e.g. narrow leaves, long roots and thick fleshy stems.

Animals **compete** with each other for scarce resources such as
space, food and water.

Plants compete with each other for space, water and nutrients from the soil.

Competition helps to **limit** the size of populations.
Other factors that affect the size of populations are:
food availability, disease, loss of habitat and predation.

Predators are animals that catch and kill their own food.
The animals that they catch are called their **prey**.

Questions

1 Copy and complete:

Many …… factors affect living things. They
have to be …… to survive these conditions.
There is a lot of …… for scarce resources.

Water, …… and …… are key resources for
…… and animals. Competition and other
factors like disease and …… limit the size of
populations.

2 Both camels and polar bears have big feet.
Why is this an advantage to both animals?
(Hint: think about how difficult it is to walk on
sand or snow.)

3 White objects lose less heat by radiation.

Use this fact to give another reason (apart from
camouflage) why polar bears are white.

4 a) Name four things (other than competition) that limit the size of a population.

    b) For each one explain how it affects the population.

5 Match up the animals and plants in column A with their correct habitat in column B.

Column A	Column B
tadpole	playing field
daisy	hedge
owl	sea shore
earthworm	river bank
seaweed	pond
salmon	woodland
hawthorn	river
otter	soil

6 Copy and complete this table showing how polar bears survive in the arctic.

Feature	How it helps survival
thick fur	
thick layer of blubber	
greasy fur	
rounded shape	
strong legs	

7 Why do flowers growing on a woodland floor flower early in the year? (Hint: look back to photosynthesis in Chapter 6.)

8 Why is it an advantage for wildebeest to stay in large herds?

9 Find out why many people are trying to ban hunting with hounds.
What is your opinion on this?

10 Copy this drawing of a growth curve.

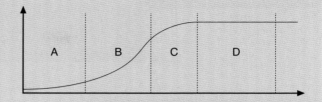

    a) Match up the letters on the diagram with these descriptions:

       i) The population size stays constant.

       ii) The population size increases slowly.

       iii) The rate of increase in population size slows down.

       iv) The population size is increasing quickly.

    b) Why does the population only increase slowly at first?

11 When wild animals are introduced to a new environment they increase rapidly.
Explain why they are able to do this.

12 a) Why are fishermen predators?

    b) Why do we not call farmers predators?

    c) Do some research on the fishing industry. How can they manage to catch so many fish that some types are nearly wiped out?

13 Describe three ways that forest fires can affect the size of an animal population.

14 Draw a mind map to summarise all you know about factors affecting populations.

CHAPTER 11

Feeding relationships

▶▶▶ 11a Food chains

In developed countries the supply of food
is a simple thing.
There are basically two groups:
the **producers** and the **consumers**.
The producers are the farmers
and we are the consumers.

In the natural world there is a similar system.

Some living things produce food from raw materials.
These are the green plants.

a) How do plants make their own food?
(Give a one word answer please.)

Animals are the consumers.
They have to eat either plants or other animals.

Consumers can be put into three groups:

- **Herbivores** – these are animals that eat plants.
  For example rabbits eat grasses and many other plants.

b) Name two other herbivores.

Making food is hard work!!!

- **Carnivores** – these are animals that eat other animals.
  For example a fox eats rabbits and other herbivores.

c) Name two other carnivores.

- **Omnivores** – these are animals that eat both plants
  and other animals.
  We are an example of an omnivore.
  Another example is the grouse.
  This is a moorland bird that feeds on heather shoots
  and small insects.

d) What is the difference between a herbivore and a
vegetarian?

Food chains are simple diagrams that show what eats what.

They are always written like this:

$$A \rightarrow B \rightarrow C$$

A is the **producer** which is eaten by
B which is the **herbivore**.
C is the **carnivore** which feeds on the herbivore.

e) There might be another link in the chain labelled **D**. What type of feeder would this be?

Food chains must always begin with the producer (green plants).

f) Why must they always begin in this way?

Here is a simple food chain:

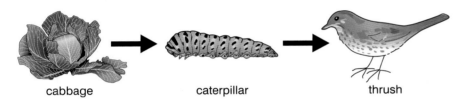

cabbage   caterpillar   thrush

The arrows show the direction in which energy is flowing.

In a food chain → means eaten by.

Food chains can also be found in water:

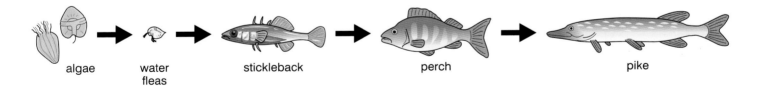

algae   water   stickleback   perch   pike
           fleas

Remind yourself!

1 Copy and complete:

Green plants are also called ...... They make food by ...... Animals have to eat their food, they are called ...... Plant eaters are called ...... and ...... eat other animals. A diagram showing what eats what is called a ....

2 Where does the initial source of energy for all food chains come from?

(Hint: look back at Chapter 6.)

3 A badger is another example of an omnivore.

Find out what kind of plants and animals the badger eats.

Have you ever counted up how many
different types of food you eat?
The answer is probably lots (although not all
of them are good for you!).
Very few animals only have one source of food.
So in the real world single food chains
do not exist.
What we have are **food webs**.

A food web is made up of lots of food chains
all linked together.

Food webs have all the same key features
as food chains:

- they start with producers (green plants)
- they contain consumers (herbivores, carnivores
  and omnivores)
- arrows show the direction of energy flow.

Here is an example of a food web from the Antarctic:

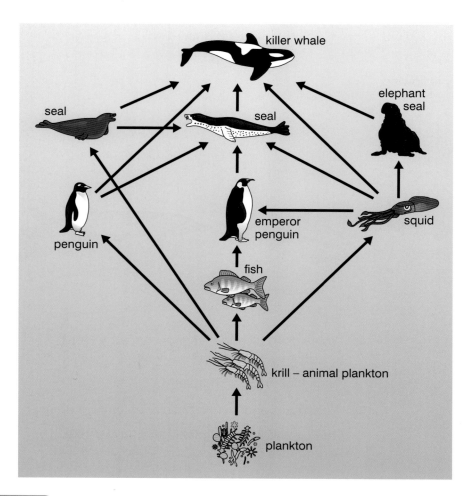

The tiny plants and animals in this web
are called **plankton**.
Plant plankton are the producers of the sea.
Animal plankton are very important because
they are the only source of food for the blue whale.

Krill – the food for blue whales.

a) What will happen to the blue whale population
if all the plankton die out?

Like many food webs the Antarctic web is quite complex.
In exam questions you are often asked questions like:

'What will happen to the killer whale population if there are
less seals available?'

Answer: There will be fewer whales because there is less food,
or the whales eat more of the squid and penguins.
Here is an example from an area of British woodland:

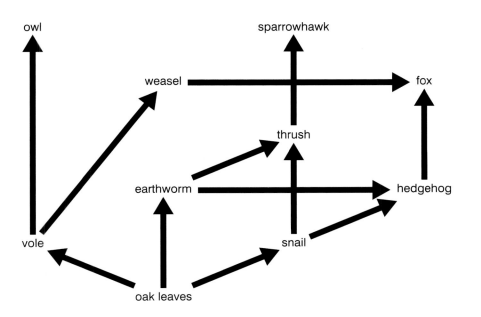

b) If all the thrushes were killed what would happen to:
  i) the number of sparrowhawks
  ii) the number of earthworms?

Remind yourself!

1 Copy and complete:

Food webs are made up of a series of ....
linked together. They exist because animals
have lots of ...... sources of food. If one source
of food ...... an animal can always switch to
something else.

2 In the food web above write out two examples of
food chains.

3 Look at the Antarctic food web.

Which animals would have to depend on other
food sources if all the squid died out?

Pyramids of number

Food chains are good for showing us
what eats what.
But they don't give us any idea
of how many organisms are involved.
It takes lots of plants to feed one herbivore,
and lots of herbivores to feed one carnivore.

Look again at this simple food chain:

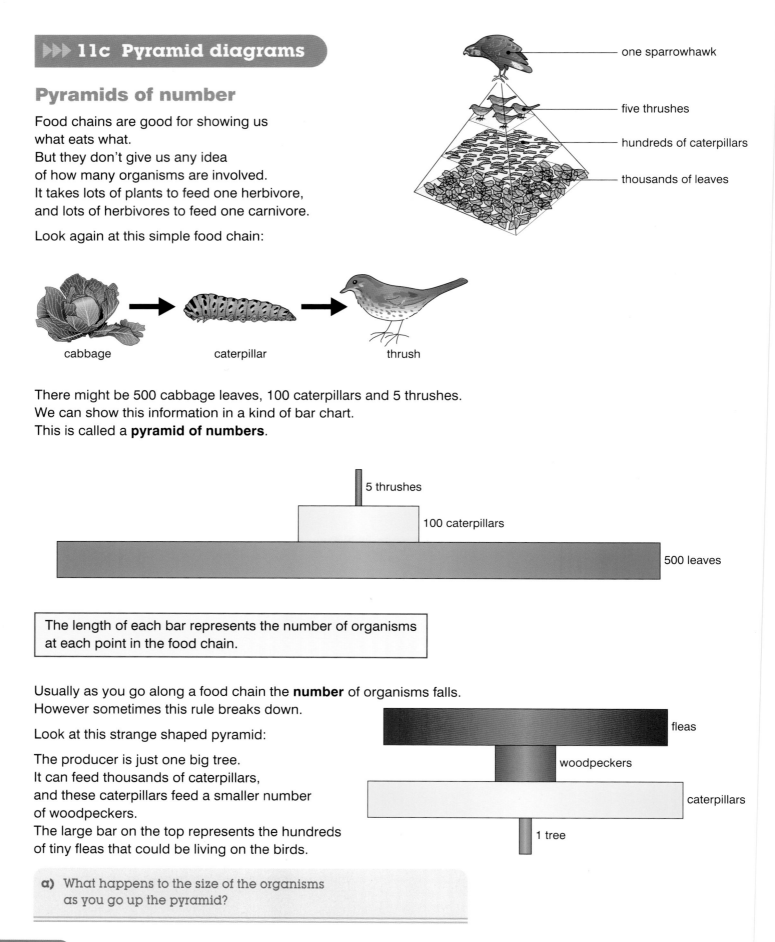

one sparrowhawk

five thrushes

hundreds of caterpillars

thousands of leaves

cabbage   caterpillar   thrush

There might be 500 cabbage leaves, 100 caterpillars and 5 thrushes.
We can show this information in a kind of bar chart.
This is called a **pyramid of numbers**.

5 thrushes

100 caterpillars

500 leaves

> The length of each bar represents the number of organisms
> at each point in the food chain.

Usually as you go along a food chain the **number** of organisms falls.
However sometimes this rule breaks down.

Look at this strange shaped pyramid:

The producer is just one big tree.
It can feed thousands of caterpillars,
and these caterpillars feed a smaller number
of woodpeckers.
The large bar on the top represents the hundreds
of tiny fleas that could be living on the birds.

fleas

woodpeckers

caterpillars

1 tree

a) What happens to the size of the organisms
as you go up the pyramid?

Pyramids of biomass

As we have just seen, pyramids of number
do not take into account how big the organisms are.
So a better way of showing this information is to use
a **pyramid of biomass**.

> **Biomass** is the mass of living material.

This time the length of the bar will represent the mass
of the total number of organisms.

> A **pyramid of biomass** shows the mass
> of all the organisms at each point in a food chain.

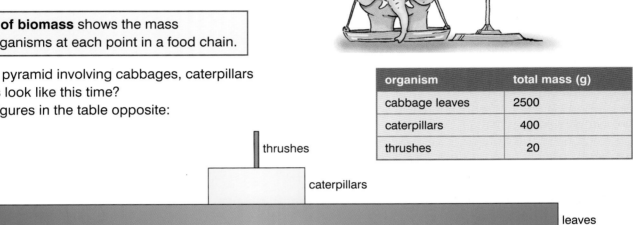

What will the pyramid involving cabbages, caterpillars
and thrushes look like this time?
Look at the figures in the table opposite:

organism	total mass (g)
cabbage leaves	2500
caterpillars	400
thrushes	20

thrushes

caterpillars

leaves

The pyramid of biomass is still a 'normal' pyramid shape.
But what about the strange shaped pyramid
involving the tree and the fleas?

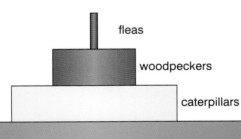

fleas

woodpeckers

caterpillars

1 tree

Unlike last time this is also a 'normal' pyramid shape.
This is because one tree will weigh many tonnes,
but even thousands of fleas will only weigh a gram or two.

Remind yourself!

1 Copy and complete:

Food chains can be shown in pyramids of ......
or ...... The length of the ...... represents either
the numbers or the biomass. Biomass is the
...... of living material.

2 Complete this sentence:

Pyramids of biomass are more useful than
pyramids of number because they take account
of ......

3 Draw and label a pyramid of number that is not a
pyramid shape.

The original source of energy for food chains
is sunlight.
This energy is transferred into chemical energy
in the leaves of green plants.
The chemical energy is stored in starch.

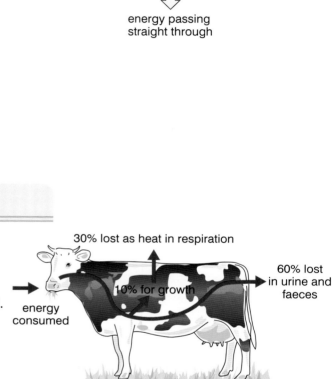

a) What part of a plant cell traps sunlight energy?

Only about 10% of the energy reaching the plant
is transferred into chemical energy.
Of the other 90% some energy:

● is reflected from the leaf
● passes straight through the leaf
● simply heats up the leaf.

> Photosynthesis is not very **efficient** at transferring
> light energy into chemical energy.

b) What does the plant use this chemical energy for?

It is not just plants that are **inefficient** at transferring energy.
Out of every 100 g of plant material available to herbivores,
only 10 g is used for growth.
In other words only 10% of the available energy is transferred.
Of the other 90% some is:

● lost in food that isn't eaten
● lost in food that isn't digested (it passes out as faeces)
● used in respiration.

c) Which parts of a plant might not be eaten?

Energy gains and losses in a herbivore.

A similar loss of energy happens at every stage in a food chain.

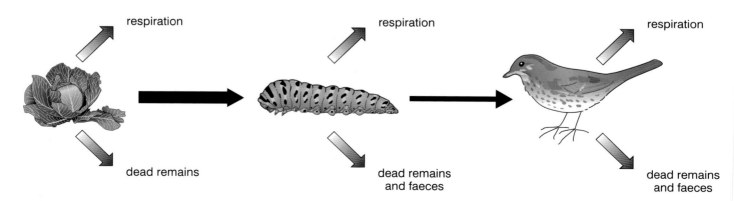

respiration   respiration   respiration

dead remains   dead remains   dead remains
                    and faeces   and faeces

Efficient feeding

As we have seen, less and less energy is available
the further we go up a food chain.
This explains why food chains are not usually
more than five links long.
In a food chain any longer than this
there would be hardly any energy left.

Look back at section 11c.

d) What happens to the number of animals
as you move up a food chain?
Explain your answer.

If we want to eat an energy-efficient diet
we should eat from low down in the food chain
(eating the plants rather than the herbivores).
This is more efficient because we have cut down
on a lot of the energy loss.

Growing plants for food.

A farm growing wheat can help to make food
that will support lots of people.
A farm growing grass to feed cattle
will provide food (meat) for far fewer people.

Vegetarians are people who choose not to eat meat.
Their diet makes the best possible use
of the energy in food chains.
Despite not eating meat, vegetarians
can still eat a balanced diet.

Meat farming provides less energy.

e) Some farms have hilly land with poor soil.
This land is unsuitable for growing crops.
What animals are usually kept on this type of farm?

Remind yourself!

1 Copy and complete:

At every stage in a food chain energy is ......
The leaves of plants ...... some of the Sun's
energy. When animals move around they use
...... At every stage, only about ......% of the
available energy is transferred. This explains
why food chains are never very ...... Feeding
...... down in the food chain makes best use of
the energy.

2 In a country where food is in short supply, why is
it better to grow wheat rather than raise cattle?

3 Most people get protein from meat. Where do
vegetarians get their protein from?

4 Find out about intensive farming of poultry.

How does this type of farming try to
cut down on energy losses in food
chains?

Summary

Food chains are diagrams showing what eats what.

The original source of energy for food chains is **sunlight**.

Food chains always begin with green plants.
Green plants are known as **producers** because they make their own food.

The other organisms in a chain are called **consumers**.

Consumers who only eat plants are called **herbivores**.
Consumers who eat other animals are called **carnivores**.
Consumers who eat both are called **omnivores**.

Food webs are lots of food chains linked together.
They are more realistic because most animals eat more than one food.

In both food chains and webs, arrows show the **transfer of energy** between one organism and the next.

Pyramid diagrams show food chains in a different way.
Pyramids of number show how many organisms there are at each level in a food chain.
Pyramids of biomass show the mass of all the creatures at each level.

Green plants only capture a small amount of the available sunlight energy.
Only a small amount of the energy available in plant material is transferred to the consumers.
A lot of energy is 'lost' in respiration and undigested food.
This is why food chains are fairly short.

To gain the most energy from food feeding should be as low down the food chain as possible.

Questions

1 Copy and complete:

Food chains always begin with ......

Producers can make their own ...... They are eaten by the ...... These organisms are made up of ...... and ...... Food webs are made up of interconnected food ...... Feeding relationships can also be shown in pyramids of ...... and ......

2 Look at this food web:

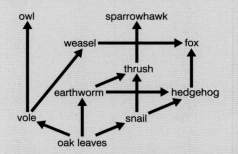

a) Name one example of:

  i) a producer

  ii) a herbivore

  iii) a carnivore.

3 Look at this seashore food web:

sea bird

crab   starfish

barnacle → dog whelk ← mussel

limpet

small algae

a) From this web write out two separate food chains.

b) Name two herbivores and one carnivore from this web.

c) Which animals in this web are predators?

d) If all the mussels died out what might happen to:

   i) the number of starfish

   ii) the number of small algae?

4 Look at the following pyramid of number diagrams:

a)

b)

c)

Now match each pyramid with one of the food chains below:

i) oak tree → bark beetles → woodpecker → fleas

ii) grass → rabbit → fox

iii) rosebush → greenfly → ladybirds → sparrows.

5 This table shows information about a food chain:

organism	numbers	mass
leaves	200	1000 g
caterpillars	100	400 g
thrushes	5	300 g
kestrel	1	200 g

a) On graph paper draw a pyramid of numbers for this food chain.

b) What do you notice about the numbers of organisms as you go up the pyramid?

c) Now draw a pyramid of biomass.

d) How does the biomass change as you go up the pyramid?

e) If the producer was a tree how would this change the shape of the pyramid of numbers.

f) Draw the shape of the pyramid.

6 a) What percentage of energy reaching a leaf is not transferred into chemical energy?

b) Describe two ways in which the rest of the energy is lost.

c) How is most of the energy consumed by a herbivore lost?

7 Find out the difference between the diet of a vegetarian and a vegan.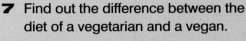

8 Many supermarkets sell fish produced on fish farms.

Find out about fish farms and write an information leaflet about them for your class.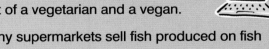

CHAPTER 12

Nutrient Cycling

▶▶▶ **12a Death and decay**

Every autumn thousands of trees lose their leaves.
These leaves blow around for a bit and tidy gardeners
sweep them up.
After a few weeks they are all gone.
The same goes for all the other dead plants.
Dead animals soon disappear too.
What have all organisms got in common?
They all rot away when they die.
In biology we call rotting away **decomposition**.

a) What type of trees lose their leaves?

Two types of organism bring about decomposition.

Detritivores are small animals that cut up
dead material into small pieces.
Some examples of detritivores are:
earthworms, woodlice and maggots.
Without these organisms decomposition
would be much slower.

The main group of **decomposers** are microbes.
Bacteria and fungi (moulds) break down
dead material using enzymes to digest it.
They then absorb the digested material
into their cells.

b) Which are smaller, detritivores or microbes?

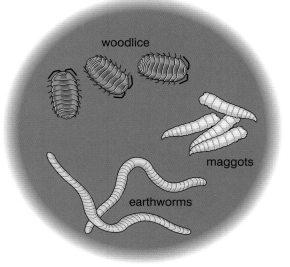

woodlice

maggots

earthworms

Some useful small animals.

- Decomposers work best in warm, damp
  conditions.
- Plenty of oxygen also helps the microbes to work.
- Oxygen isn't essential though, and often the
  smelliest decay occurs where oxygen is lacking.

c) Why do dead things hardly rot at all in the Arctic
region?

*The fungus is growing in the dead tree
trunk and decomposing it.*

Why is decomposition important?

In Chapter 6 we saw how plants need nutrients.
Plant roots absorb mineral salts including nitrate
needed for healthy growth.

d) Where do plants get these nutrients from?

There is not an endless supply of these nutrients.
They need to be **recycled** and this is where
decomposers come in.
The diagram opposite shows how nutrients
are recycled.

- When dead organisms rot away these
  nutrients are released into the soil.
- Plants will then absorb them through their
  roots, and use them for growth.
- The plants are then eaten by animals.
- Then the nutrients will be passed
  further along the food chain.
- When these plants and animals die
  they will rot away.
  So the whole **cycle** begins again.

Although food chains always begin with producers,
these do not need to be living.

Here is an example of a **decomposer food chain**.

Dead leaves → fungus → beetle → frog

Here is an example of a **detritivore food chain**.

Dead animal → maggots → blackbird → sparrow hawk

You will notice that this food chain appears to begin with a
consumer. However, a dead animal cannot eat anything and here
acts as a producer!

*Eventually all the leaves will have rotted
away.*

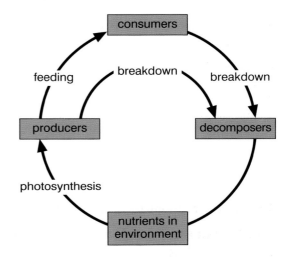

Remind yourself!

1 Copy and complete:

When living things die they …… Decomposition
is helped by animals like …… and ……
Microbes like …… and …… are the main
decomposers. Decomposition releases useful
…… back into the ……

2 What are the best conditions for decomposition?

3 Why does more decomposition happen in
summer than in winter?

4 Make a list of things that do not rot away.

For many years people have put decomposition to good use.
We rely on microbes to rot away a lot of waste.
All the food material that goes in your dustbin ends up on rubbish tips.
Once there it rots away and adds nutrients to the soil.

Composting

For hundreds of years man has used decomposition to make garden soil more **fertile**.

A lot of decomposition goes on here.

a) What is meant by fertile soil?

> A **compost heap** is a place that speeds up natural decomposition.

The things that can be put on compost heaps are:
- garden waste such as grass cuttings
- kitchen waste such as potato peelings
- shredded newspaper.

Eventually all this material will rot away.
What you will be left with is **compost**.
It looks a bit like soil with bits of fibre in it.
Compost makes excellent fertiliser for your garden.
It is a good way of replacing nutrients in the soil.

To make good compost you need:
- warmth
- moisture
- plenty of air (for oxygen), and
- lots of detritivores.

Very useful decomposition.

To encourage these conditions many gardeners use a compost bin.
This is like a dustbin with holes in the side.
The holes allow air and small animals in.
The lid allows heat to build up and it also keeps moisture in.

b) What material should not be put in a compost bin?

lid keeps heat in and rain and animals out

air holes to allow air to circulate

open base to allow microbes to get in

Sewage treatment

Another good example of useful decomposition is sewage treatment.

> Sewage is the mixture of urine, faeces and washing up water from homes and industry.

It is not safe to dump raw sewage in rivers. It contains lots of microbes that could spread disease.

Efficient sewage treatment did not begin until the 1920s.
Before then many rivers were heavily contaminated with raw sewage.

Most sewage now goes to special treatment works.
- Here it is filtered to remove big solid material like wood and plastic.
- Grit and stones that have been washed off roads are then allowed to **settle** out in big tanks.
- The solid faeces is also allowed to settle out. Eventually this will be used as fertiliser.

Now microbes can get to work.
- After settling and separating the remaining liquid is called **effluent**.
- A variety of microbes break down the effluent into simpler substances.
- These substances are usually harmless **minerals**, **gases** and **water**.
- There is a final period of settling which removes the microbes.
- The water that is left is clean enough to be pumped into a river or the sea.

Blackpool beach used to be badly affected by sewage, but now there is a new treatment plant.

At a sewage works microbes break down harmful waste.

Remind yourself!

1 Copy and complete:

...... is a very useful process. Man uses microbes to make ...... for the garden. This uses up kitchen and ...... waste and creates ...... for the soil. Decomposition also helps to make ...... sewage safe.

2 Why is it not safe to dump raw sewage in rivers?

3 Why should you not put weeds that have been sprayed with weed killer in a compost bin?

▶▶▶ 12c The carbon cycle

Carbon is one of the most important elements
for living things.
They use it to make carbohydrates, proteins,
fats and lots of other molecules too.

a) Which gas in the air contains carbon?

Plants take in carbon dioxide through their leaves.
They use it in photosynthesis to make food.
Animals get their carbon by eating plants
or other animals.

There isn't very much carbon dioxide in the air.
So why doesn't it run out?
Like lots of other useful chemicals it is
recycled.

Remember the word equation for respiration?

oxygen + sugar → energy + carbon dioxide + water

When plants and animals respire they return
carbon dioxide to the air.
The same thing happens when decomposers
feed on dead plants and animals.
They respire too.

Human activity also returns carbon dioxide to the air.
Not only by respiring but also by burning
fossil fuels.

b) Name some fossil fuels.

Fossil fuels contain a lot of carbon.
They were formed millions of years ago
from dead plants and animals.

As a result of all these processes
there is never a shortage of carbon dioxide.
In fact as we will read in Chapter 13
too much carbon dioxide can be a big problem.

c) Look at the carbon cycle at the top of the page.
Which is the slowest part of the cycle?

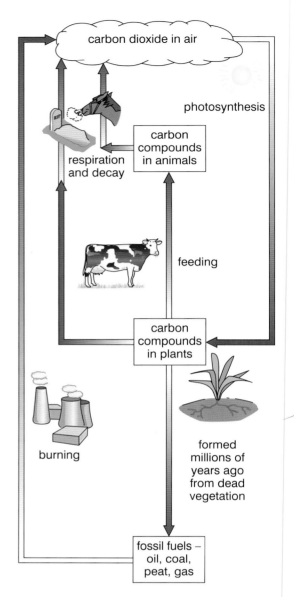

The carbon cycle.

*Humans put a lot of extra carbon dioxide
into the air.*

Summary

Dead plants and animals are broken down by **bacteria and fungi**.
These microbes are called **decomposers**.
They are helped by small animals called **detritivores**.

Decomposers work best in warm, damp conditions with plenty of oxygen.
People have made use of decomposition in a number of ways.
For example, compost heaps and sewage treatment works.

The most important use of decomposition is the recycling of nutrients.
When plants and animals rot away these chemicals are returned to the soil.
They can then be used again first by plants and then by animals.

Carbon is needed by all living things.
It is used to help build the bodies of plants and animals.
It is found in the air as carbon dioxide.
This gas is removed by plants for use in photosynthesis.
It is returned to the air by respiration and the burning of fossil fuels.

Questions

1 Copy and complete:

Bacteria and ......break down dead plants and
...... These ...... work best when it is warm
and ...... Decomposition returns useful
chemicals to the ...... Plants can then absorb
these chemicals through their ...... Man uses
microbes to produce ...... for the garden and
to clean up ...... One of the most important
elements is ...... This is also ......
Photosynthesis, ... and burning play a big
part in recycling carbon.

2 a)   Explain why dead material rots more
quickly in warm damp conditions.

b)   How do small animals help the rotting
process?

3 If nutrient recycling is so good, why do farmers
need to add fertilisers to their land? (Hint: do
farmers let crops rot away on their land?)

4 Look at this diagram of the carbon cycle:

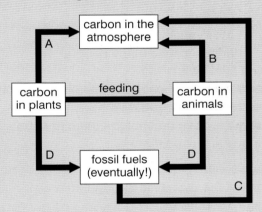

a)   Copy the diagram and use these words to
label processes A to D.

**burning   respiration   decay
photosynthesis**

b)   Which process could add lots of extra
carbon dioxide to the air?

c)   Why might chopping down large areas of
rain forest increase the amount of carbon
dioxide in the air?

CHAPTER 13

▶▶▶ 13a Taking over the land

Does your town have a bypass?
Bypasses are designed to take traffic
away from town centres.
This makes the town a more pleasant place.
But what about the surrounding countryside?

a) What damage could a new road do to an area
of countryside?

b) What are the advantages to a town of having
a bypass?

In the early 1990s Newbury in Berkshire
got its long-awaited bypass.
It had taken a long time to build.
This was partly because many people
protested against it.
Protesters set up camps along the proposed route
and then refused to move.
Eventually the government took them to court
and the police moved them on.

Why did so many people protest?
The route of the bypass is through an area
of attractive countryside.

Bypasses often lead to strong protests.

People were afraid that the bypass would:
- destroy the habitats of animals
- kill rare species of plant
- spoil the countryside as a place for walkers
- lead to more traffic and therefore more pollution.

When bypasses are proposed, planners have a difficult
decision to make.
On the one hand, there is damage to the countryside.
On the other, there are the benefits of a traffic-free town.
Quite often a compromise is reached.
The bypass is approved, but its route is changed.
The new route is usually through less sensitive countryside.

Would you build a road through here?

Other uses of land

Bypasses are only one example of how
man uses up land.
Land that is home to many plants and animals.

Building houses

Lots of people prefer to live in country areas.
This means that towns are spreading more
and more into the countryside.

c) Why do many people prefer to live in the countryside?

Farming

Our ever increasing population needs more
and more food.
Farmland takes up a lot of space and this has
a big effect on animals and plants.

d) How does chopping down hedgerows
help farmers to grow more food?

Another new housing estate.

Dumping waste

The more people there are the more waste they make.
Rubbish tips also take up a lot of space, as well as
being very smelly.

e) What is meant by the term biodegradable rubbish?

Quarrying

Quarries produce stone for new houses and roads.
Some quarries are right in the heart
of attractive countryside.

f) What can be done with deep quarries
when there is no stone left?

Quarrying uses up a lot of land.

Remind yourself!

1 Copy and complete:

When man uses up land he affects both ......
and ......

The ...... of animals are destroyed. Also many
rare ...... of plant can be killed. The greater the
size of the ...... the more land is used up.

2 How do quarries lead to more pollution from
traffic?

3 Make a list of the benefits to the public of traffic-
free town centres?

▶▶▶ 13b Population problems

Do you know the size of the world's population?

> By the year 2000 it was over 6000 million,
> and it is increasing by around 20 000 a day!

Obviously the population hasn't always been
this big.
Look at this graph which shows how the human
population has grown:

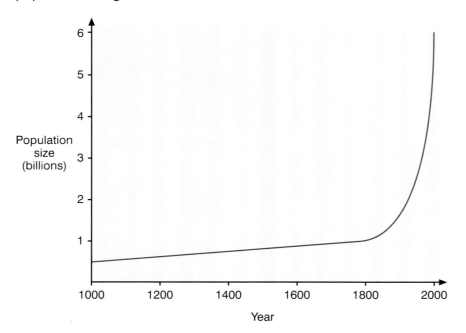

For many years the human population
only increased very slowly.
A lack of food and poor medical care meant that
people did not live very long.

As well as food shortages and disease, many wild populations
are controlled by predators.

a) Is the human population controlled by predators?
Explain your answer.

b) From the graph when did the human population start to
grow rapidly?

There are a number of reasons for this increase:
- improved treatment of disease
- more and better food
- cleaner water supplies.

c) How does having clean water help a population to grow?

Does man have any predators?

What are the consequences of rapid population growth?

Man's impact on the environment increases with the size of the population.

More people means:

- More land is needed for homes.
- More food is needed so more land is turned over to farming, and more chemicals are used.
- More waste is produced so more waste disposal sites are needed.
- More pollution of the air, land and water.

There are a number of things that can be done to reduce the damaging effects of population growth:

- More homes can be built on **brownfield** sites. These are areas often in towns which were once occupied by industry.
- The amount of packaging around foods in supermarkets can be reduced to cut down on waste.
- Using more renewable energy sources will cut down on pollution.

The human population is ever increasing.

d) Give an example of a renewable energy source.

Given that people are living for longer it makes sense to reduce the birth rate.
This ought to slow down the population increase.
However having large families is important in poorer countries.
So reducing the population growth this way is not easy.

e) What is meant by a brownfield site?

f) Why is building on these sites preferable to building in the countryside?

New homes could be built here rather than on countryside.

Remind yourself!

1 Copy and complete:

The human population is growing very ......
Better food, medical care and cleaner ......
supplies mean that people live longer. More
people means a greater ...... on the
environment.

2 Why are large families important in poor countries?

3 Find out how China has tackled the problem of a rapidly increasing population.

Look at this photograph of a northern town
in the 1940s.
Smoke from coal fires is pouring out
of most of the chimneys.
Today it would be almost impossible
to find a similar scene.

In 1956 the **Clean Air Act** started a process
that has greatly reduced air pollution in the UK.
Today even those homes that burn solid fuel
such as coal use 'smokeless' varieties.

The Clean Air Act was a relief to big towns.

a) What problems would people have experienced
before the Clean Air Act?

> **Pollution** means anything that damages the environment
> and the living things in it.

The most common cause of air pollution
is the burning of **fossil fuels**.
- Fossil fuels like coal and oil all contain **carbon**.
- When they burn they give off **carbon dioxide**.
- They also contain a variety of other chemicals,
  such as **sulphur**.
- When fossil fuels are burned this sulphur
  forms **sulphur dioxide** gas.

b) Which gas does sulphur react with when it burns?

Sulphur dioxide is produced naturally by volcanoes.
But humans produce over **100 million tonnes** every
year by burning fossil fuels.
Sulphur dioxide damages our breathing system.
In the famous London smogs of 1952
sulphur dioxide and smoke caused 12 000 deaths.

Due to changes in industry sulphur dioxide pollution
is now less of a problem in big cities.
But we are still living with the problems that were
created when smogs were an everyday event.

A natural polluter.

Acid rain

You will know from your chemistry lessons
that we measure acidity on the pH scale.

Fortunately rain is not this acidic!

---

c) What number on the scale is neutral?

Most people think that rain is neutral.
In fact normal rain is slightly acidic.
This is because it has carbon dioxide dissolved in it.
Its pH number is about 5.6.
Fortunately this is not acid enough to harm living things.

When sulphur dioxide and nitrogen oxides
dissolve in rain water they make it much more acidic.
Now it will be pH 4.0 or less!
This is **acid rain**.

Acid rain damages plants.
- In the Black Forest in Germany many trees
  have lost all their leaves.
- Acid rain also makes lakes and rivers acidic.
- This causes the death of animal plankton.

---

d) What is animal plankton?
(Hint: see Chapter 11.)

- These invertebrates are important food for fish.
- The acid water also affects the gills of fish.
  This means that many fish die of suffocation.

The effects of acid rain have built up
over many years.
In Europe the worst affected area is Scandinavia.
This is because most of the wind blows from the
southwest.
So most of the pollution is carried on the wind
and then falls as acid rain over this region.

*This water looks very clean but it is
probably very acidic.*

Remind yourself!

1 Copy and complete:

Pollution is anything that ...... the environment
and the ...... things in it. Most air pollution
results from burning ...... fuels. These create
gases like carbon ...... and ...... of sulphur.
Sulphur oxides cause damage to our ......
system.

2 Why isn't rainwater neutral?

3 Why is air pollution less of a problem today?

4 How does the death of tiny invertebrates affect
the numbers of fish?

▶▶▶ 13d The greenhouse effect

Carbon dioxide is not as harmful to humans as sulphur dioxide.
However it is helping to create a big pollution problem.

Carbon dioxide is known as a **greenhouse gas**.
This is because like the glass of a greenhouse it helps to keep heat in.
It traps heat in the Earth's atmosphere.

a) Where does the heat come from originally?

Methane is another greenhouse gas.
- Some methane is released from waste tips when rubbish decays.
- It also comes naturally from the rice fields of Asia. Here the land is very marshy and this is ideal for making methane.
- Cattle also produce a lot of methane – from both ends!

The amount of both of these gases is slowly rising.
- Carbon dioxide levels are rising as we burn more and more fossil fuels.
- Another big source of carbon dioxide is **deforestation**.
  In tropical regions large areas of forest are cut down to provide wood, and land for farming.
- Taking away the trees means that *less* photosynthesis takes place.
- Less photosynthesis means *less* carbon dioxide is removed from the air.
- To make matters worse the trees are often not simply chopped down.
  Large areas are burned down, and this gives off even more carbon dioxide.

Methane levels are rising due to the need to produce more food for the increasing population.

The greenhouse effect.

Deforestation is a big source of pollution.

Why worry about carbon dioxide?

Scientists believe that greenhouse gases
may be causing **global warming**.
This means that all over the world
average temperatures may rise.
Europe for example could be 2–3°C warmer
by the year 2030.

If you like sun bathing this sounds terrific.
- But a rise in temperature could melt
  some of the polar ice caps.
  This could cause the sea level to rise.
  and lead to flooding of low lying areas.
- Also the rise in temperature could affect
  the world's weather patterns.
  Areas that currently grow a lot of the world's
  crops might gain a desert climate.
  If this happens then food production could
  be badly affected.
  Even without global warming, the world
  struggles to feed its population!

b) Which country in Europe might suffer most
if sea levels rise?

*Global warming might make this ice
disappear.*

This farm could become an arid desert.

Cleaning the air

There are many things we can do
to prevent air pollution, such as:
- Factory chimneys can be fitted with **scrubbers**.
  These are devices that remove harmful gases
  like sulphur dioxide from smoke.
- Encouraging the use of public transport.
  If there are less cars on the road then less
  fossil fuels will be burned.

c) Now add your own suggestions to this list.

I'm doing my bit for the environment.

Remind yourself!

1 Copy and complete:

Carbon dioxide and …… trap …… in the Earth's
atmosphere. This is causing …… warming. If
…… gases are not ……, world food production
might be affected. Also …… levels will rise
causing ……

2 Explain fully how deforestation affects carbon
dioxide levels?

3 Wheat is an important basic food source.

Find out which regions produce most
of the world's wheat.

Look at this photograph.
It shows how oil pollution can seriously harm wildlife.
Once released into water oil can spread very quickly.
In 1991 Iraq invaded Kuwait.
During the Gulf war millions of litres of oil
were deliberately released into the sea.
Many seabirds died as a result of this pollution.

Oil spills do great harm.

a) What is the usual cause of oil pollution?

- Oil poisons sea birds and clogs up their feathers.
- The oil also kills many plants and animals living on the seashore.
- Oil pollution is very difficult to remove.
- Some of the detergents used can also kill marine creatures.

Oil is the most common pollutant in the sea.
In lakes and rivers there are lots of other nasty chemicals.

- Untreated sewage and the waste from farmyards together with fertilisers from farms can badly affect rivers.
- Apart from possibly containing poisonous chemicals, sewage and farm waste contain **nitrates**.

Nitrates are minerals needed by plants for growth.
But what happens if too much nitrate gets into a river?

- Water plants (**algae**) grow rapidly;
- they cover the surface and block out the light;
- lack of light means the water plants below die;
- these dead plants are decomposed by bacteria;
- the bacteria use up all the oxygen;
- a lack of oxygen causes fish and other animals to die.

The results of too many chemicals.

As a result the river becomes **stagnant** (smelly and lifeless).

This type of pollution is called **eutrophication**.

It is much worse in slow flowing rivers or lakes.
In fast flowing water the pollution is quickly carried away.

Other farm chemicals

Other farm chemicals can also pollute water.
Farmers add chemicals to their crops to kill
pests that damage them.

> Chemicals used to kill pests are called **pesticides**.

b) Give one example of an animal that destroys crops.

These chemicals can be very dangerous in water.
Sometimes they are washed off the land by heavy rain.
Or the farmer might be careless when spraying.

In a lake these chemicals can work their way
up a food chain.
As each creature eats lots of the organism below it,
the concentration of the pesticide increases.
Eventually the concentration will be high enough to
kill.
Usually it is the animal at the top of the chain
that dies first.

Pesticides have also been known to kill land animals too.
For example birds of prey like eagles
have been poisoned by pesticides.
These pesticides were originally used on seeds.

c) How could this pesticide have affected the eagles?

Nowadays there are strict regulations on the
use of pesticides.
● Chemicals that do not break down naturally are banned.
● **Biological control** uses a natural predator of the pest.
  These predators are bred in large numbers.
  They are then released and start to eat the pest.

d) How do we remove pesticides from food
before we eat it?

Locusts are a big pest in many countries.

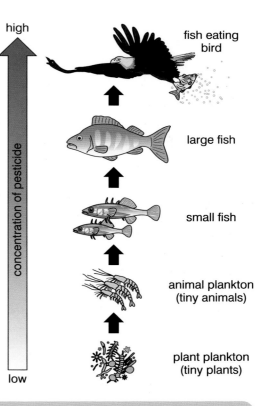

Remind yourself!

1 Copy and complete:

Water is polluted by a variety of chemicals
including …… and the waste from …… Sewage
leads to rivers becoming …… They run out of
…… and all the animals die. Some farm
chemicals like …… get washed into rivers and
build up in food ……

2 Farmers add nitrates to the soil. How can these
chemicals end up in rivers?

3 Find out about a pesticide called DDT.

Why is it banned?

Our impact on the environment is
threatening the survival of many plants and animals.
Not just exotic creatures like the panda,
but more familiar ones like cod.
The world's fish stocks are being threatened
by **overfishing**.
New technology means that single boats
can catch many tonnes of fish in one go.

A commercial fishing trawler.

> **a)** What effect will a fish shortage have
> on the price of fish and chips?

Deforestation is another example of how humans
are threatening other species.
Rain forests are the habitat for many animals
that are found nowhere else on earth.
Many rain forest plants are a source
of powerful medical drugs.

The forests are cleared to provide land for:
- hardwood for building
- other types of wood for fuel and paper
- farming e.g. cattle and rubber trees.

Clearing the forests leads to soil erosion.
Also, as we have already seen, it may
contribute to the greenhouse effect (see page 142).

> **b)** What is the quickest way for humans
> to clear an area of forest?

> **Conservation** means preserving and protecting
> habitats such as forests and oceans.

Obviously we cannot expect people to stop
using the Earth's natural resources.
The key to good conservation is **sustainable
development.**

> **Sustainable development** means using
> the Earth's resources without destroying
> them in the process.

Some good examples of sustainable development are:

- reducing the numbers of fish caught
- catching a bigger variety of fish
- replacing each tree cut down with a new one
- using biological control to kill pests instead of pesticides
- finding alternatives to fossil fuels
- recycling waste.

In Britain we have developed **National parks**. The idea is to give legal protection to natural habitats but also to allow the public to enjoy the countryside. Many of these parks contain **Sites of Special Scientific Interest** (**SSSIs**).
These often contain plants or animals whose numbers are very low.
These are known as **endangered species**.

A number of endangered species are being bred in zoos.
The idea is to build up a stock of these animals.
When numbers are large enough they can be released back into the wild.

What can ordinary people like us do to help conservation?

- Walk to the shops instead of going by car;
- take glass and paper to recycling bins;
- turn off lights when leaving a room;
- use recycled paper;
- put kitchen and garden waste into a compost bin.

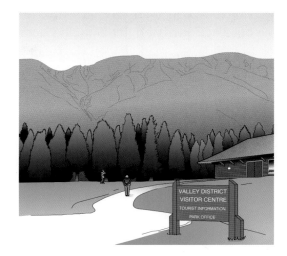

> **c)** Read through this chapter again.
> In what other ways could you protect the environment?

A contribution to conservation.

Remind yourself!

1 Copy and complete:

...... means protecting and preserving the environment. Without conservation many ...... and ...... species will disappear. Cod stocks are low because of ...... To protect them we need to use ...... development.

2 Why is it important to protect plants that live in rain forests?

3 What are alternative sources of energy to fossil fuels?

4 Find out which animals in the world are most endangered

Summary

Humans reduce the land available for other species by:
building, quarrying, farming and dumping waste.

They also **pollute** the air, water and land with a variety of chemicals.

Human impact on the environment is directly linked to the **population size**.
As the population increases so does the damage caused by humans.
The human population is growing rapidly due to better medical care, more food and improved water supplies.

Burning **fossil fuels** releases gases such as **sulphur dioxide**.
This causes **acid rain**.
Acid rain kills both plants and animals.
Air pollution also damages our breathing systems.

Burning fossil fuels also releases **carbon dioxide**.
Carbon dioxide traps heat in the atmosphere causing **global warming**.
Global warming might cause sea levels to rise and changes in the world's climate.
Methane from cattle and rice fields also contributes to the greenhouse effect.
Deforestation also affects the amount of carbon dioxide in the air.
Less trees mean less carbon dioxide is removed by photosynthesis.
Also, burning forests adds carbon dioxide to the air.

Water is polluted with a variety of chemicals, including:
oil, pesticides, fertilisers and raw sewage.

Conservation means preserving and protecting the environment.
Protecting the Earth's natural resources can best be done using
sustainable development.
This means using resources in a way that does not destroy them for future generations.

Questions

1 Copy and complete:

Humans have a big …… on the environment.
This impact is directly linked to the …… of the population. The more people there are, the more …… occurs. Humans pollute the ……, ……, and land with a variety of …… By using sustainable …… we can protect the environment for future ……

2 a) Which two gases cause the greenhouse effect?

b) Where do these gases come from?

c) How do they cause global warming?

d) Explain how global warming might cause floods in low lying areas.

3 Describe two ways that zoos can help to conserve wildlife.

4 Find out what the initials CITES stand for.

5 The African elephant is the world's largest land mammal, and in recent years its numbers have fallen dramatically.

   a) Find out what has caused the population to fall.

   b) What has been done to protect the elephant?

6 a) What does the word pollution mean?

   b) Write down some examples of how humans pollute the Earth.

7 a) Explain why normal rainfall is slightly acidic.

   b) What gases dissolve in clouds to form acid rain?

   c) What effect does acid rain have on

     i) trees and

     ii) fish

   d) What effect has acid rain had on buildings in old industrial towns?

   e) What can humans do to reduce acid rain pollution?

8 Find out about CFC gases.

   a) Where do they come from?

   b) What damage do they do?

   c) How are humans tackling this pollution problem?

9 Some parts of Canada have been affected by acid rain originating in the USA.
Explain how pollution from one country can affect another country.

10 What uses could an old quarry be put to when it has run out of stone?

11 In connection with building what is the difference between a 'greenfield' site and a 'brownfield' site?

12 Why is a lot of rubbish buried rather than being burnt?

13 a) What items of household rubbish can be recycled?

   b) If more rubbish was recycled how would this affect the amount of land needed for rubbish tips?

14 What is the best way of ensuring that there will always be enough trees to supply our need for wood?

15 The table below shows the amounts of methane given off from various sources:

Source of methane	Percentage(%)
farm animals	19
marshes	27
rice fields	23
rubbish tips	11
mining	15
oceans	5

   a) Draw a bar chart of these figures.

   b) Which source contributes most gas?

   c) Which two sources are not influenced by humans?

   d) Apart from being a greenhouse gas, how else is methane a problem?

16 Draw a mind map to summarise all you know about the impact of humans on the environment.

Further questions on the Environment

▶ **Habits**

1 Beavers live in the streams of northern forests. They cut down lots of trees. They use some trees for food and some to make dams. The dams form large deep ponds. In the middle of these ponds the beavers build their homes. These homes have underwater entrances.

   (a) Explain how each of the following features makes beavers well adapted for living in water and cutting down trees:

   (i)   large paddle-shaped tails

   (ii) thick, waterproof fur

   (iii) front teeth which never stop growing   (3)

   (b) Suggest **two** ways in which the design of the beaver's home protects it from predators.   (2)

   (AQA (NEAB) 1999)

2 • Common seals are large mammals which live on our coasts.
   • They feed on fish which they catch underwater.
   • Their bodies are long and streamlined, with smooth skin.
   • They have a thick layer of fat under their skin.
   • The back legs are modified to form flippers.
   • They are able to swim in cold water for long periods.
   • The patterns on their skin makes them camouflaged.

   (a) What are animals which catch and eat other animals called?   (1)

   (b) Why is it an advantage for the seals to be camouflaged?   (1)

   (c) Suggest reasons for seals having:

   (i)   streamlined bodies with smooth skin;   (1)

   (ii) a thick layer of fat under their skin;   (1)

   (iii) legs modified to flippers.   (1)

   (AQA (NEAB) 2000)

3 Cactus plants grow in areas which receive very little rain.

   (a) How does each of the following adaptations help these plants to survive in these conditions?

   (i)   The outside of the stem is covered with a thick layer of wax.   (1)

   (ii) The leaves have been reduced to long, sharp thorns.   (1)

   (b) The stems of these plants are green because their outer cells contain a green pigment.

   (i)   Name this green pigment.   (1)

   (ii) Explain why it is an advantage to these plants to have stems with cells containing the green pigment.   (2)

   (AQA 2001)

4 The graph shows how a population of bacteria in a muddy puddle changed over a period of four days.

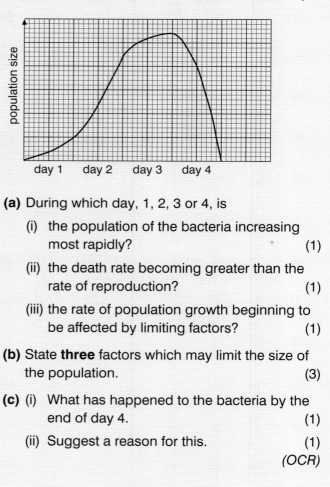

   (a) During which day, 1, 2, 3 or 4, is

   (i)   the population of the bacteria increasing most rapidly?   (1)

   (ii) the death rate becoming greater than the rate of reproduction?   (1)

   (iii) the rate of population growth beginning to be affected by limiting factors?   (1)

   (b) State **three** factors which may limit the size of the population.   (3)

   (c) (i)   What has happened to the bacteria by the end of day 4.   (1)

   (ii) Suggest a reason for this.   (1)

   (OCR)

Further questions on the Environment

▶ **Feeding relationships**

5 Sand eels are small fish found in the North Sea. The diagram shows the energy transfers between these fish and other organisms.

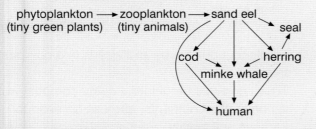

(a) What do we call this type of diagram? (1)

(b) Phytoplankton provide energy for all the animals.
Explain how phytoplankton can be energy suppliers. (3)

(c) Humans are removing large numbers of cod and herring from the North Sea.
Some people say that this will increase the number of sand eels, others say that the number will decrease.
Both groups of people could be correct.
Explain how:
Sand eels might increase
Sand eels might decrease. (4)
(AQA 2001)

6 Several animals live in a National Park. The table shows what they eat.

Animal	Food
bighorn sheep	green plants
elk	green plants
marmots (small mammals)	green plants
mountain lions	bighorn sheep, elk, snowshoe hares
snowshoe hares	green plants
wolves	elk, marmots, mountain lions

(a) Copy and complete the food web for the animals listed in the table.

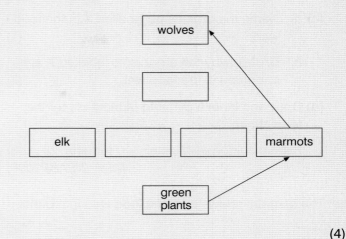

(4)

(b) Many tiny fleas live in the fur of the wolf.
Draw a pyramid of numbers for this food chain.
green plants → marmots → wolves → fleas (2)
(AQA (NEAB) 1999)

7 The diagram shows a food web. All the organisms except humans and sea birds live in sea water. Plankton are microscopic organisms.

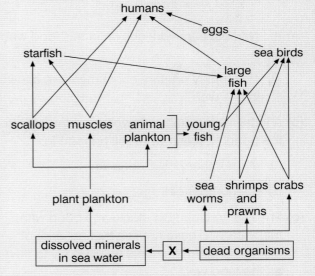

(a) **(i)** Name the producers in this food web. (1)

(ii) What is the importance of producers in this food web? (1)

(b) One food chain in this food web is:
Dead organisms → Sea worms → Large fish → Humans
Describe what this food chain tells us. (3)

(c) (i) Name the organisms labelled **X** in the food web. (1)

(ii) Explain why the organisms labelled **X** are important in the food web. (2)

(d) Draw and label a pyramid of biomass for the food chain:

Plant plankton → Mussels → Humans (2)

(AQA (NEAB) 2000)

8 The diagram shows part of a food web in a pond.

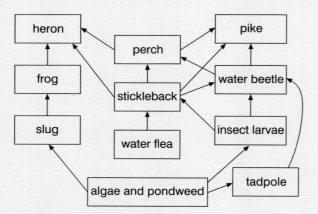

(a) (i) Name a carnivore shown in this food web. (1)

(ii) How many primary consumers are shown in this web? (1)

(b) A local fishing club removes all of the pike from the pond. Explain what will happen to:

(i) the number of sticklebacks (1)

(ii) the number of frogs. (2)

(c) The fishing club now stocks the pond with many carp. These are fish that eat a lot of plants. Explain the effect this is likely to have on the food web. (4)

(EDEX)

▶ **Nutrient cycling**

9 The constant cycling of carbon is called the carbon cycle.
Describe in detail how carbon is cycled by living grass plants. (4)

(AQA (NEAB) 2000)

10 Gardeners put leaves and weeds in a compost heap.
They leave this to rot into a peaty compost which is rich in plant nutrients, and can be used to improve soil fertility.

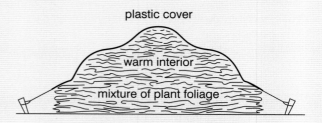

Three gardeners living in different parts of Britain made compost heaps on the same day in Spring.
To speed the rotting process they covered them with plastic sheets and mixed them up each week using a fork.
The table gives some information about where the gardeners lived.

Gardener	Town	Average temperature /°C	Average annual rainfall /cm
A	London	13.2	61.0
B	Manchester	12.0	85.9
C	Edinburgh	11.0	69.9

(a) (i) Suggest which gardener's compost heap will be ready first. (1)

(ii) Explain your answer. (1)

(b) How does mixing up the heap help speed up rotting? (1)

(c) The gardener in Manchester forgot to replace the plastic sheet over his compost heap. Explain how this might affect the rotting. (1)

(OCR)

11 The diagram shows some of the stages by which materials are cycled in living organisms.

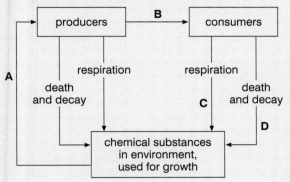

(a) In which of the stages, **A**, **B**, **C** or **D**:

(i) are substances broken down by microbes;

(ii) is carbon dioxide made into sugar;

(iii) are plants eaten by animals? (3)

(b) In an experiment, samples of soil were put into four beakers. A dead leaf was put onto the soil in each beaker. The soil was kept in the conditions shown.

Beaker **W**	Beaker **X**	Beaker **Y**	Beaker **Z**
Warm and wet	Cold and wet	Cold and dry	Warm and dry

In which beaker, **W**, **X**, **Y** or **Z**, would the dead leaf decay quickest? (1)

(AQA (NEAB) 1999)

▶ **Humans and the environment**

12 Some large forest areas are being destroyed. This changes the amount of carbon dioxide in the atmosphere.

(a) (i) State **one** use for the trees that are cut down. (1)

(ii) State **one** use for the cleared land. (1)

(iii) How has the destruction of forests affected the amount of carbon dioxide in the atmosphere? (1)

(b) (i) How has the destruction of forests caused an increased greenhouse effect? (4)

(ii) State **one** effect of an increase in the greenhouse effect. (1)

(AQA (SEG) 1998)

13 This question is about pollution.

(a) Copy and complete this passage using these words. You may use each word once or not at all.

cars	dissolve	evaporate	fuels
kill	plants	soot	sulphur	water

Fossil ...... burnt by industry and ...... can release ...... dioxide into the atmosphere. This can ...... in ...... to form acid rain. When this falls it can ...... fish and damage ....(7)

(b) Carbon dioxide is produced by many industries.

(i) Name **two** types of environmental problems that a build up of carbon dioxide could cause. (2)

(ii) Apart from industry, how could carbon dioxide build up in the atmosphere? (1)

(AQA (SEG) 2000)

14 (a) Give **three** different ways in which humans reduce the amount of land for wild animals. (3)

(b) Name **three** materials which may pollute rivers. (3)

(c) Describe, in as much detail as you can, the processes which result in the formation of acid rain. (4)

(AQA (NEAB) 1999)

15 People pollute the atmosphere with gases produced by power stations, factories and engines. Some of these gases form acid rain. The sentences **A–D** are in the wrong order. Write them in the correct order in a flow diagram like this.

A The gases dissolve in rain.

B Fossil fuels are burned.

C Lakes and rivers become acidic so plants and animals die.

D Sulphur dioxide and nitrogen oxides are released.

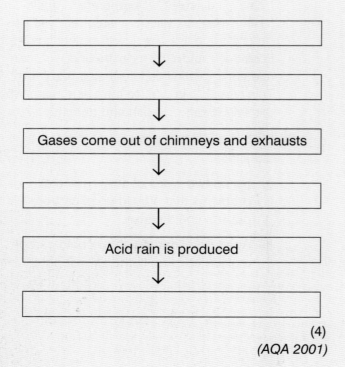

(4)
(AQA 2001)

16 Lichens are simple plants. The table shows how many different types of lichen were recorded at set distances from a city centre.

Distance from city centre in km	Number of types of lichen found in a given area
0	4
2	7
3	10
5	20
6	25
7	40

(a) Draw a graph of these results with distance on the X axis. (4)

(b) Use your graph to estimate the number of types of lichen at 4 km from the city centre. (1)

(c) Use information from your graph to describe how the number of types of lichen is linked to the distance from the city centre. (1)

(d) Lichens are killed by air pollution. Name **two** gases that pollute the atmosphere in a city. (2)

(AQA (NEAB) 1998)

17 Between 1880 and 1980 it has been estimated that about 40% of all tropical rainforest was destroyed; much of it was destroyed by burning.

(a) Give **three** reasons why this large-scale deforestation has happened. (3)

(b) Large-scale deforestation is affecting the levels of carbon dioxide in the air.
  (i) What is happening to carbon dioxide levels? (1)
  (ii) How does deforestation cause a change in the carbon dioxide levels. (2)
  (iii) Explain some long-term effects that deforestation is likely to have on climate and soil fertility. (5)

(c) Many types of habitat are being destroyed, with serious effects on the wildlife living there. What methods can scientists and governments use to protect this wildlife? (2)

(EDEX)

Section Four
Inheritance and selection

In this section you will find out what makes members of the same species different from each other.
You will also learn about how man can create animals and plants with particular features.
You will learn about the inheritance of sex and about inherited diseases.
You will discover how living organisms have gradually changed over millions of years.

CHAPTER **14**   **Variation**

CHAPTER **15**   **Inheritance**

CHAPTER **16**   **Reproduction**

CHAPTER **17**   **Selective breeding**

CHAPTER **18**   **Evolution**

VARIATION

▶▶▶ 14a Spot the difference

If you are using this book in school take a look
around the classroom.
You will notice that everyone has a number of
different features or **characteristics**.
Even a pair of identical twins will have some
differences you can spot.

a) Look at two of your friends.
Write down three differences
between them.

These differences are important because
they are between members of the same species.

> A **species** is a group of plants or animals
> of the same kind.

For example, all humans belong to the same species.

b) What is the proper name for the human species?

Members of the same species have a lot of features
in common.
But they also have a number of differences.

*How many different species can you see
here?*

> **Variation** describes the differences between members
> of the same species.

There are two causes of variation in plants and animals.

Genetic variation is caused by different people **inheriting**
different **genes**.
Genes are small 'packets' of information passed on from
parents to their children.
This information is found in the nucleus of cells.
It controls all of your characteristics.
People in the same family look similar to each other.
This is because they share lots of the same genes.

Families are similar but not identical.

c) Why do people from different families look very different to each other?

d) Make a list of human features that result from genetic variation.

Unrelated people have fewer similarities.

Not all our differences can be put down to our genes.
If a new pupil joins your class from another area
they might have a different accent.
You are not born with an accent you develop it.
It is affected by the way your parents, friends
and other local people talk.
In other words it is caused by your environment.

Environmental variation results from all the things
that influence you as you go through life.
Environmental variations are often very obvious,
e.g. your hair style or your fashion sense (or not!).

Some features result from your genes
and your environment.
Both your parents may be very slim.
But if you stuff yourself with chips everyday
you're quite likely to end up overweight.

e) Can you think of any other features that might be both genetic and environmental?

The Slim family and their son Big Jim.

Plants also show variation.
Rose bushes for example have lots of different
coloured flowers.
The colours are the result of inherited genes.
Different bushes may grow to different heights.
This difference will probably be environmental.
It will be to do with the amount of light,
or the amount of nutrients in the soil.

f) Hydrangea plants can have blue or pink flowers. What affects the colour of the flowers?

Remind yourself!

1 Copy and complete:

Plants and animals belong to different ...... All the members of one species will be similar but not the ...... The differences are called ...... Some variation is caused by ...... and some by the ......

2 What is meant by the terms:

a) genetic variation

b) environmental variation?

3 What accent will develop in children born in America but with English parents?

Explain your answer.

157

Have you heard the expression
'It's in his genes'?
When someone says this they usually
mean something that is inherited.
But what exactly are genes, and where
in the body are they?

Genes are small sections of bigger structures
called **chromosomes**.
Chromosomes are found in the nucleus
of cells.

A pair of chromosomes.

a) What is the job of the nucleus?
(Hint: look back to Chapter 1.)

Chromosomes come in pairs and in each human
body cell there are 23 of these pairs.

b) How many chromosomes does each cell contain?

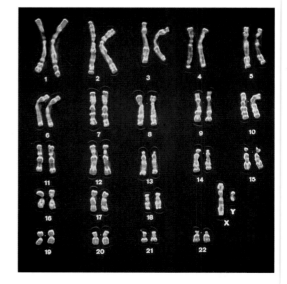

What do you mean I've only got four pairs of chromosomes?

Not all species have the same number of chromosomes.
A cat only has 19 pairs, and a simple fruit fly
only 4 pairs.

Chromosomes are made of a very special chemical.
This is called **deoxyribonucleic acid**.
It is usually called **DNA** for short.

Look at the photograph opposite:
It shows a human **karyotype**.
This is a display of all the chromosomes from
a normal body cell.

Notice that each pair is different.
They gradually get smaller and they are not all
the same shape.

Producing a chart like this is one way that doctors
can check for signs of faulty chromosomes.
Faulty chromosomes can cause disorders like
Down's syndrome.
This is caused when there is just one extra chromosome.

c) What does DNA stand for?

d) Name a disorder that is caused by
faulty chromosomes.

Sex cells are different

Sex cells (**gametes**) have fewer chromosomes.
Instead of 23 **pairs** they just have 23 **single** chromosomes.

To make a new human being the two sex cells have to join together.
This happens inside the female and is called **fertilisation**.

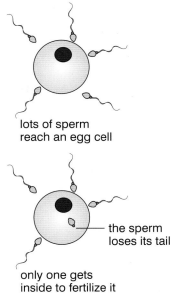

lots of sperm
reach an egg cell

the sperm
loses its tail

only one gets
inside to fertilize it

Fertilisation.

> **e)** If the sperm and the egg each had
> *23 pairs* of chromosomes, how many pairs would the new
> individual have?
> (Hint: it's double what it should have.)

But because both sex cells have just 23 single chromosomes
the new individual ends up with 23 pairs.
Exactly the right number!

Genes on chromosomes

Each chromosome contains hundreds of genes.

> Each gene carries the information for a particular characteristic.

For example, there are different genes for hair, eye
and skin colour.
Because chromosomes come in pairs, then genes
also come in pairs.
So in fact each characteristic is controlled by a
pair of genes.
In each pair of genes the information might be slightly
different.
For example, one of the pair of genes for hair colour
might be for blond while the other is brown.

> Genes in a pair that carry different information
> are called **alleles**.

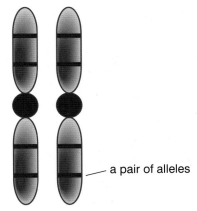

a pair of alleles

The dark bands show the position of genes.

Remind yourself!

1 Copy and complete:

Chromosomes are found in the …… of a cell.
They are made of …… In most human cells
there are …… pairs. In sex cells there are only
23 …… chromosomes. Each chromosome
holds hundreds of ……

2 What is an allele?

3 Find out about National 'Jeans for genes' day.

Write a report about this campaign.

Look at how small 1 mm is on your ruler.
Some cells are 100× smaller than this.
Imagine a cell this small containing something
2.5 m in length!

a) How can something so long fit into a
microscopic cell?

This is exactly the situation with DNA.
As you have already seen DNA is the chemical that
chromosomes are made of.
It is found in the nucleus of all our cells.

DNA is often called the blueprint of life.
In the days before computers, engineers used to
draw up plans on large sheets of blue paper.
Not surprisingly these plans became known as
'blueprints'.

All the DNA in a cell contains the complete design
for the organism that it is found in.
Every cell carries an identical copy of this plan.

We know that genes are small sections of
chromosomes.
Therefore if chromosomes are made of DNA,
genes must be short pieces of DNA.

Molecules of DNA are twisted into a shape called
a 'double helix'.
This is just like a ladder that has been twisted.

b) Think of another word that describes a helix.

The discovery of the structure of DNA was a major
event in the history of science.
Most of the work was done by a group of
scientists at Cambridge University in the 1950s.

These scientists cracked the code.
In other words they worked out how a chemical
could carry the instructions for all our features.

c) Why were scientists excited when they
discovered the structure of DNA?

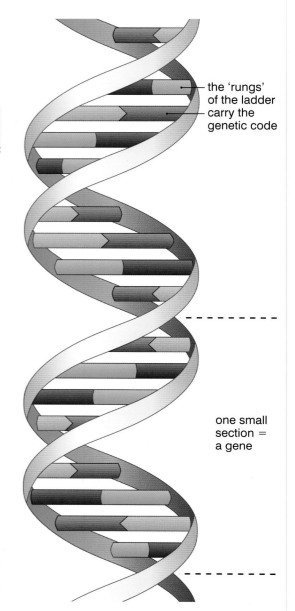

the 'rungs'
of the ladder
carry the
genetic code

one small
section =
a gene

The DNA double helix.

Copying the code

Every time a cell divides to make two new ones
the DNA must be copied.
This makes sure that every cell has all
the instructions it needs to work properly.
The DNA helix unwinds itself, is copied and then
winds back up again.

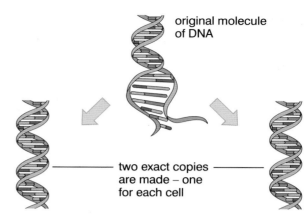

original molecule
of DNA

two exact copies
are made – one
for each cell

DNA can copy itself.

The Human Genome project

Ever since the structure of DNA was worked out
scientists have wanted to identify each separate gene.
This is not an easy task.
It is estimated that there are about 30 000 genes
on the 23 chromosomes.
In 1990 the Human Genome project began.

The **human genome** is all the genes in each cell.

By 2001 the project was virtually complete.
Scientists knew which genes were found on which
chromosomes.
This knowledge will be especially valuable in
treating **genetic disorders**.
These are problems caused by faulty genes,
e.g. Huntington's disease and cystic fibrosis.

> **d)** Why did the human genome project take so
> long?

Scientists can now identify which genes cause
these disorders.
Having done this, they now stand a better
chance of curing the problem.
One possible solution is to replace the faulty gene
with a new one.
This is called **gene therapy** and at the moment it
is still at the experimental stage.
But one day, who knows?

Studying DNA has many potential benefits.

Remind yourself!

1 Copy and complete:

DNA carries the complete …… for a living
organism. It is contained in the …… of each cell.
It is a very …… molecule wound up into a ……
helix.

2 What is meant by the term 'The human
genome'?

3 Everybody has different DNA. How is this
information useful to the police?

Look at the photograph opposite.
It shows a man with a condition called
albinism.

a) How does this man look different to a typical
male?

This man is an **albino** and the condition is caused by a
mutation.

A **mutation** is a change in a gene or chromosome.

People with albinism are missing a special
chemical in the skin called **melanin**.
Melanin gives our skin its colour and is responsible for suntan.
Melanin also protects our skin from harmful ultraviolet
light.

Melanin is produced in the skin.
The instructions for melanin production are carried
by a gene.
In albinos this gene has been changed and so melanin
is not produced.

b) Why do people with this condition have
to avoid strong sunlight?

This condition also occurs in wild animals.
This photograph shows an albino hedgehog:

c) Why might it be a disadvantage for a hedgehog
to be white?

You probably haven't seen an albino in real life.
This is because albinism is very rare.
In Europe and North America less than 5 people in
every 100 000 are affected.

Fortunately, although many mutations are
harmful, they are also relatively rare.

An albino male.

*People from Africa have more melanin than
Europeans.*

An albino hedgehog.

How are mutations caused?

Mutations can just happen naturally.
We call these mutations **spontaneous**.
But we do know that certain things make mutations much more likely.

- Radiation such as X-rays and gamma rays can cause mutations.
  Have you ever had an X-ray?
  If so you might have noticed that the radiographer does not stay in the room while it is being taken.

> **d)** Why do you think this is?
> (Bear in mind how many X-rays the radiographer will take each day!)

- Nuclear bomb blasts or leaks from nuclear power stations can cause mutations.
  Following the Second World War lots of Japanese babies were born with mutations.
  This followed the use of atom bombs in 1945.
- Lots of chemicals are known to cause mutations.
  For example, tar in tobacco smoke.
  Chemicals in tobacco smoke can cause cancer.

Cancer occurs when cell division gets out of control.
This is because the gene that switches off cell division has been changed.
Cancer causes large growths of cells called tumours.
Sometimes tumours are harmless but when they are **malignant** they are very difficult to treat.
This photograph shows a diseased lung removed in an operation.

> **e)** What is the main cause of lung cancer?

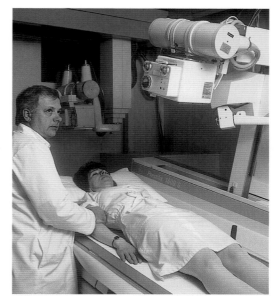

An occasional X-ray will not cause a mutation.

A cancerous lung.

Remind yourself!

1 Copy and complete:

A change in a gene is called a ...... These changes often occur ...... However ...... and certain ...... make them more likely. If the gene that switches off cell division changes then ...... occurs.

2 Why is melanin very important to our skin?

3 Find out about the 'Slip, Slap, Slop' campaign in Australia.

Gregor Mendel was an Austrian monk born in 1822. He combined a job growing food for the monks with an interest in how characteristics are inherited.

- Mendel took two varieties of pea plants, tall and short.
- He cross bred these plants and collected the seeds that developed.
- When he grew new plants from these seeds they all grew **tall**.

> Mendel concluded that the tall plants must carry an instruction that over-rules the short plants.

- Surprised by this, he then cross bred two of these tall plants.
- Again he collected and planted the seed. This time he found that $\frac{3}{4}$ of the plants were **tall** and $\frac{1}{4}$ **short**.

> From these results Mendel concluded that these instructions must come in pairs.

The tall plants from his first experiment must have inherited both a tall and a short instruction. However only the tall instruction showed itself.

In the second experiment some of the plants inherited two short instructions. This is why they were short and not tall.

> Although he didn't realise it Mendel was in fact talking about **genes**.

The importance of Mendel's work was not realised until after his death in 1884. This was partly because it was first published in a small scientific report that not many scientists read. Also Mendel used a lot of clever statistics in his work, and at the time this was not well understood by biologists.

Gregor Mendel.

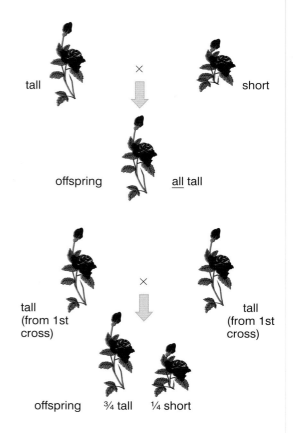

tall × short

offspring   all tall

tall (from 1st cross) × tall (from 1st cross)

offspring   ¾ tall   ¼ short

Summary

Animals or plants of the same kind are members of the same **species**.

Different members of a species have different **characteristics**.
These differences are called **variation**.
Variation has two causes:
Genetic variation is caused by the **genes** an individual inherits.
Environmental variation is caused by the conditions in which they develop.
Sometimes variation is caused by genes *and* the environment.

Genes are small sections of **chromosomes**.
Each gene carries the instructions for a particular characteristic.
Chromosomes are made of a chemical called **DNA**.
They are found in **pairs** in the nucleus of a cell.
In all human body cells there are 23 **pairs** but in sex cells there are 23 **single** chromosomes.

Like chromosomes, genes are also found in pairs.
Each one of a pair of genes is called an **allele**.
Each allele will carry slightly different instructions,
e.g. one for blond hair and the other for black hair.

Mutations are changes to genes and chromosomes.
They can occur naturally but are often caused by radiation and chemicals.
Most mutations are harmful and can lead to conditions like cancer.

Gregor Mendel was the first scientist to work out the way characteristics are inherited.

Questions

1 Copy and complete:

...... carry the instructions for all our characteristics. They are found on ...... in the nucleus of a cell. Chromosomes occur in ...... except in ...... cells where they appear singly. Chromosomes and genes are made of a chemical called ...... Members of the same ...... have different ...... and this is called ...... Variation is caused by ...... that are inherited and also by the ...... Some characteristics such as a person's ...... can be affected by genes and the environment.

2 Why do you think Mendel did not know about genes or chromosomes in 1866?

3 Why is it necessary for sex cells to only have half the normal number of chromosomes?

4 Why is sunbathing without any protection dangerous?

5 Why do crop plants near trees not grow as well as those in the middle of a field?

6 If it works why will gene therapy be an important medical treatment?

CHAPTER 15 INHERITANCE

▶▶▶ 15a Chromosomes and sex

Look around your class and count up
the number of boys and girls.
Are there about equal numbers of each?
If so this shouldn't be a surprise.
In the human population the ratio of
males to females is about 50:50.

The reason for this lies in the way sex is inherited.

Look at the photograph opposite.
It shows a pair of chromosomes from a male.

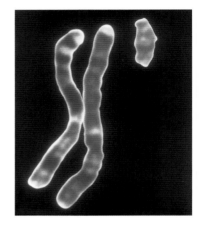

a) What difference can you see between them?

This is the 23rd pair of male chromosomes.
They are called the **sex chromosomes**.
The large one is called the **X** chromosome
and the small, the **Y** chromosome.

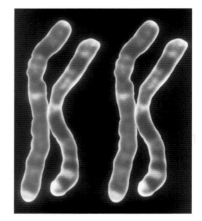

> Sex chromosomes carry the genes that
> make us male or female.

Now look at this photograph.
It shows the female sex chromosomes (again the 23rd pair).

b) What do you notice about these chromosomes?

The really important difference between the male
and female is the presence of the Y chromosome.
In 1990 scientists discovered the male gene.
Guess where they found it?
That's right, on the Y chromosome!

c) Why does the Y chromosome only hold a
small number of genes?

I may be small but I'm very powerful

How is sex inherited?

We learned in Chapter 14 that sex cells have half the number of chromosomes of normal body cells.

- When sex cells are made, the pairs of chromosomes divide into two.
- This means that half of the sperm cells produced will contain an X chromosome and half a Y chromosome.
- When females produce eggs each will only contain one X chromosome.

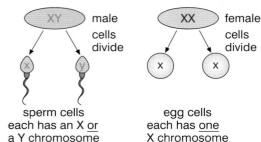

d) Why do the eggs only contain an X chromosome?

To create a new human being one egg must join with one sperm cell.

e) What do we call this process?

The **genetic diagram** below uses this information to explain the inheritance of sex.

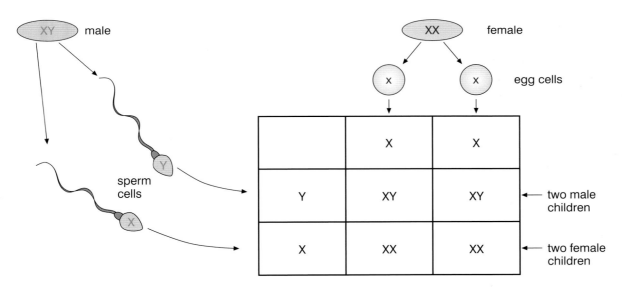

Have you ever noticed how hair colour
varies in families?
Brothers and sisters do not always share
the same colour.
In fact children can have different coloured
hair than both their parents.

a) How many different hair colours are there
in your family?

In Chapter 14 we learned that characteristics
are controlled by genes.
Hair colour is a good example of this.

On one of our 23 pairs of chromosomes
there will be a gene that controls hair colour.
But, remember that genes come in pairs
called **alleles**.

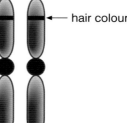
← hair colour gene

b) What is an allele?

*A pair of chromosomes containing the hair
colour gene.*

Everybody has two alleles for hair colour.
These alleles might be both the same,
e.g. they might both carry the instruction
for black hair or blond hair.
In this case you will **either** have black or blond hair.

However, the alleles might be different.
One might carry the instructions for black
and the other for blond.

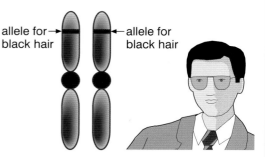
allele for → black hair
← allele for black hair

c) What hair colour will you have this time?

Well it certainly won't be a mixture.
In fact you will have black hair.
Why?
Because black is an example of a **dominant**
characteristic.

In every pair of alleles if one is dominant
that characteristic will always show up.

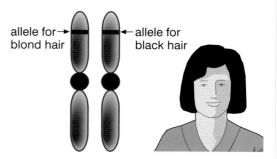
allele for → blond hair
← allele for black hair

Blond hair is called a **recessive** characteristic.

> Recessive characteristics only show up
> if **both** alleles are recessive.

Let's say that B is the symbol for the black allele,
and b is the symbol for the blond allele.

Blond is recessive.

Example 1
The father is blond and the
mother has black hair.

> **d)** What do you notice about the
> children's hair colour?

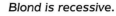

	B	B
b	Bb	Bb
b	Bb	Bb

All of the children will have
these two alleles B and b

Example 2
The father and mother both have
black hair, but they each have a
blond allele.

> **e)** Explain how it is possible for this
> couple to have a blond haired child?

	B	b
b	Bb	bb
B	BB	Bb

← this child
will be
blond
haired

There is a 1 in 4 chance of this
couple having a blond haired child

Try a third example for yourself.
This time the father has black hair but
he also has a blond allele.
The mother is blond.
Work out the example in the same way as 1 and 2.

> **f)** What are the chances of this couple
> having a blond haired child?

Remind yourself!

1 Copy and complete:

Some characteristics are …… and some are
…… Recessive characteristics only show up if
both …… are recessive. If a dominant allele is
present that characteristic will always show up.

2 Brown eyes are dominant to blue eyes.

Draw a **genetic diagram** (like the ones shown
above) to show how two brown eyed parents
could have a blue eyed child.

3 Tongue rolling is a dominant characteristic. How
many people in your class can roll their tongues?

Some diseases are not caught, like flu
but are inherited from your parents.
They usually result from a mutation.

a) What is a mutation?

Quite often the allele that is affected is recessive.
This means that to develop the disease
you must have two of these alleles.
One allele must come from each of your parents.
Fortunately this is a pretty unlikely event.
So many inherited diseases are rare.

Cystic fibrosis (CF)

Cystic fibrosis is a disease of the lungs and
the digestive system.
- Both these parts of the body normally
  contain sticky **mucus**.
- In people with this disease the mucus is
  much thicker than normal.
- This mucus blocks up the airways and tubes
  in the gut.
- Patients have difficulty breathing and absorbing food.
- The blocked airways can easily become infected.
- These infections damage the lungs and make
  breathing even more difficult.

Cystic fibrosis affects about 1 in 2000 children.
At the moment there is no cure for it.

This disease is caused by a **recessive allele**.
If both parents carry one allele, they do not
have the disease but we call them **carriers**.

The diagram opposite shows how this couple
could produce a child with cystic fibrosis.

b) What are the chances of a child inheriting CF?

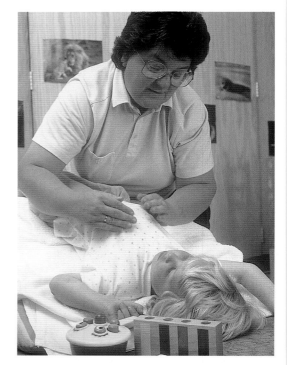

*Vigorous physiotherapy is needed by
CF sufferers.*

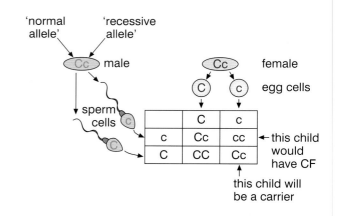

Huntington's disease

This is a disease of the nervous system.
It happens when brain cells start to break down.
The symptoms of this disease are:

- clumsy, jerky movements
- moods and depression
- memory loss
- becoming totally disabled.

This is also an inherited disease.
But this time it is caused by a **dominant allele**.
This means that you only need to inherit
the allele from one parent, not both.

The diagram opposite shows the inheritance
of this disease.

Unfortunately, Huntington's disease does not show
itself until a patient is around 40 years old.
This creates a big problem.
By the time a person is 40 they may well have
had children.
If so they may well have passed on the Huntington's allele
without even knowing it!

Fortunately, Huntington's is a rare disease.
It only affects 1 in 20 000 people.
New tests have been invented so that young people
can find out if they have the faulty allele.
This allows them to decide whether or not to risk
having children.

Huntington's disease affects people over 40.

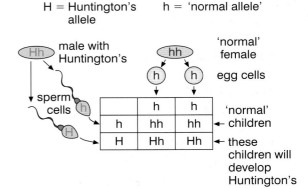

Being a carer

Diseases like Huntington's can put a lot of demands
on other members of the family.
There are the **physical** demands of helping with
feeding and dressing the sufferer.
Also there are the **emotional** demands caused by
watching a loved one suffer.
Quite often these demands can make
the carers themselves quite ill.

Remind yourself!

1 Copy and complete:

Cystic fibrosis is caused by a ...... allele. It must
be passed on by ...... parents. Huntington's
disease is caused by a ...... allele. It can be
inherited from just ...... parent.

2 What does it mean if you are a carrier of cystic fibrosis?

3 At what age do the symptoms of Huntington's disease appear?

Why is this a problem?

▶▶▶ 15d Sickle cell anaemia

Sickle cell anaemia is another inherited disease.

In Chapter 4 we looked at red blood cells.

> **a)** What shape are normal red blood cells?

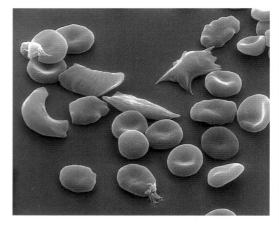

Normal and sickle shaped red blood cells.

People with sickle cell anaemia have unusual red blood cells.
Look at the photograph opposite:
It shows normal and sickle shaped red blood cells.
The sickle cells don't flow smoothly through small blood vessels.
They can cause blockages in capillaries.
These blockages can cause painful swellings, and can lead to death at a young age.

Another problem is that sickle cells don't carry oxygen as well as normal red blood cells.
As a result people with this disease get tired very easily.

Like cystic fibrosis, sickle cell anaemia is caused by a **recessive allele**.
So to get the disease a person needs to inherit this allele from both parents.

Look at the genetic diagram opposite.
It shows the inheritance of sickle cell anaemia.
In this example both the father and mother are **carriers** of the disease.

S stands for the normal allele
s stands for the recessive allele

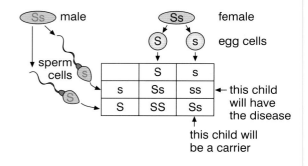

> **b)** What are the chances of this couple having a child with sickle cell disease?
>
> **c)** In terms of inheritance what is the main difference between sickle cell anaemia and Huntington's disease?

Sickle cell anaemia is not common in Britain.
If a doctor tells you that you are anaemic he usually means that you have a shortage of red blood cells.
To treat this you will usually be given tablets or medicine containing iron.

Sickle cell disease and malaria

Inherited diseases are usually quite rare.
This is true for sickle cell anaemia.
However in parts of Africa sickle cell is quite common.

Look at the map opposite:
It shows those parts of Africa where sickle cell anaemia is quite common.

Now look at this second map of Africa:
It shows the areas where the disease **malaria** is common.

■ = areas where sickle cell anaemia is most common

> **d)** What do you notice about these two areas?

■ = areas where malaria is present

Malaria is caused by a tiny single-celled creature called a **parasite**.
The parasite can be passed to humans by the bite of a **mosquito**.
It then invades the red blood cells and reproduces.
Malaria results in severe fevers and can be fatal.

- The parasite only invades **normal red blood cells**, not sickle shaped ones.
- In parts of Africa it is an advantage to be a carrier of sickle cell anaemia.
- If these people are infected by the parasite their red cells become sickle shaped.
- This gives them protection against malaria. These people are now more likely to live and have children.
- As a result sickle cell anaemia is passed on from generation to generation.

The condition survives because of the protection it gives against malaria.

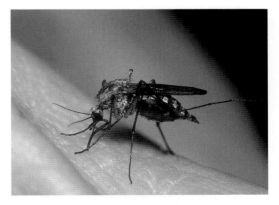

Feeding time for the mosquito!

Remind yourself!

1 Copy and complete:

Sickle cell anaemia is caused by a ...... allele.
Sickle shaped ...... blood cells block ...... They also cannot carry much ...... In parts of ...... this disease is quite common.

2 Explain why it is an advantage to be a carrier of sickle cell anaemia in parts of Africa.

3 Why do people with sickle cell anaemia often get tired?

How would you feel if you were told
you carried a genetic disease?
Would you want to have children and
risk passing it on?
What if you or your partner were pregnant,
would you want to continue with the pregnancy?

These are difficult questions that a number of people
have to face every day.

With modern technology more and more
inherited diseases can be diagnosed early.

It is the job of a **genetic counsellor** to
help answer people's questions.

A genetic counsellor has a difficult job.

The counsellor will study a family's medical history.
They will then explain:
- the chances of a child inheriting a particular disease
- how the child might be affected
- what options the parents have.

It is then left to the parents to decide what to do next.

A genetic counsellor would be able to explain
the cause of **Down's syndrome**.
- This disease is caused by a child having one
  extra chromosome.
- It can be spotted using the kind of chromosome
  chart we met in Chapter 14.
- The photograph opposite shows the
  chromosomes from a Down's syndrome child.

Look carefully:

a) Which 'pair' of chromosomes is in fact three?

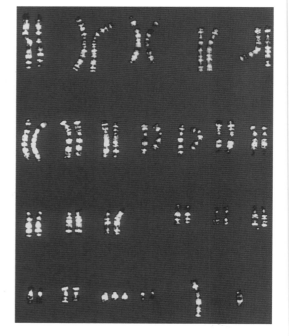

This extra chromosome gives a child physical and
mental difficulties.

b) Discuss with a friend how you would
feel if you knew your unborn child had
a genetic disease.

c) If tests were available for all genetic
diseases would you want them or not?
Explain your answer.

Summary

In humans, sex is controlled by the sex chromosomes.
Males have an **X** and a **Y** chromosome.
Females have two **X** chromosomes.
The 'male' gene is carried on the **Y** chromosome.
Half a man's sperm will carry an X chromosome and
half will carry a Y chromosome.

Many characteristics are controlled by pairs of alleles.
Some alleles are **dominant**.
The characteristic they control shows up even if that allele
is only found on **one** of a pair of chromosomes.
Some alleles are **recessive**.
The characteristics shown by these alleles only show up
if we find the alleles are found on **both** of a pair of chromosomes.

Some diseases are **inherited**.
Cystic fibrosis and **sickle cell anaemia** are caused
by **recessive** alleles.
In countries where malaria is a problem, sickle cell anaemia
gives some protection against this disease.
Both parents must be **carriers** of the allele if a child
is to develop the disease.
Huntington's disease is caused by a **dominant** allele.
Only **one** parent need have this allele for the disease to be passed on.

Questions

1 Copy and complete:

Many characteristics are controlled by ...... of
alleles. Some alleles are ...... and they always
show themselves. Other alleles are ......
These alleles will only show up if they are
present on ...... chromosomes of a pair. Some
diseases are ...... A disease caused by a ......
allele must be passed on by ...... Both parents
are known as ...... If a disease is caused by a
...... allele it can be passed on by only ......
parent.

2 Explain why people cannot be **carriers** of
Huntington's disease.

3 Why are people with CF more likely to develop
chest infections?

4 Copy and complete this genetic diagram to
show how sex is inherited in humans:

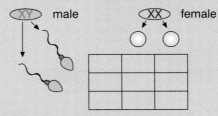

5 Not all animals have sex chromosomes.

Find out how sex is determined in
crocodiles.

CHAPTER 16

Reproduction

▶▶▶ 16a Asexual reproduction

Reproduction means producing new individuals.
Have you ever seen someone taking plant cuttings?
This is a way of growing new plants cheaply.
It involves taking a piece of stem from the original plant
and planting it in some compost.
After a while a new plant will grow.
Taking cuttings is an example of asexual reproduction.

> **Asexual reproduction** involves only one parent,
> and no joining of sex cells.

These are the kinds of organisms that reproduce asexually:

- many plants
- microbes like yeast and bacteria
- single-celled organisms like amoeba
- simple animals, such as Hydra.

Hydra is a tiny pond-dwelling animal.
It reproduces by growing a new animal
as a 'bud' on its side.
When this bud has developed enough it breaks off
and grows into an adult.

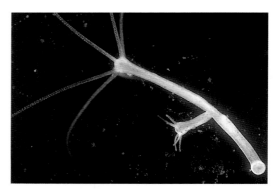

This simple animal is producing a bud.

Yeast also reproduces by 'budding'.
Eventually the bud separates from its 'parent'.

Amoeba reproduce by splitting into two.
Before this can happen, the nucleus must
divide into two.

a) Why must the nucleus divide into two?

This process takes about an hour and at the
end there are two identical amoeba.

Bacteria also reproduce by splitting into two.
Under ideal conditions this can happen for each cell once
every 20 minutes.

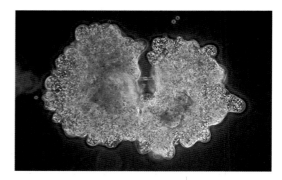

This amoeba is dividing into two.

Asexual reproduction in plants

Lots of plants can reproduce both asexually and sexually.
The asexual method is very useful if there aren't any other plants nearby.

Potatoes are really swollen stems called tubers. New shoots grow from these tubers and produce a new potato plant.

Strawberry plants have side branches called runners. These grow over the soil and produce buds. These buds produce roots and grow into new plants.

Look at this photograph of a houseplant called **Bryophyllum**.

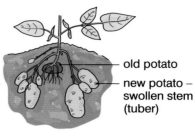

old potato

new potato –
swollen stem
(tuber)

> **b)** Where are the new plants being made?
>
> **c)** How do they end up as separate plants?

What's special about asexual reproduction?

Only one parent is involved therefore only one set of genes is involved.

> Asexual reproduction produces individuals who have identical genetic information to the parent.

> Genetically identical individuals are called **clones**.

> **d)** What do we call the type of reproduction that only involves one parent?
>
> **e)** What can you say about the genes in the offspring compared with the parent in this type of reproduction?

Remind yourself!

1 Copy and complete:

Many ......-celled organisms reproduce ......
This process often involves simply ...... into two.
Some organisms grow new ...... that eventually separate from the parent. Asexual reproduction produces ...... identical individuals called ......

2 Explain why strawberry plants produced from *one* parent do not necessarily look the same.

(Hint: look back at Chapter 14.)

3 Why are offspring of asexual reproduction **genetically identical**?

▶▶▶ 16b Sexual reproduction

Most animals and plants can reproduce **sexually**.

Sexual reproduction involves the joining together of male and female **gametes** (sex cells).

In animals the sex cells are the **sperm** and the **egg**.

a) Which is the male gamete and which is the female?

A sperm egg.

In Chapter 14 gametes were mentioned in connection with inheritance.
What makes sperm and eggs special?

Sperm cells are the only animal cell that can swim.
They have a long tail and look a bit like tadpoles.
They have a large nucleus that is packed full of the chromosomes from the male parent.

b) In humans how many chromosomes would each sperm be carrying?

Egg cells are bigger than sperm cells.
Each one contains a large nucleus.

c) What will this nucleus contain?

An egg cell.

There is also a lot of food stored in the egg.
This will be needed if the egg is fertilised.

Fertilisation occurs when a sperm and an egg **fuse** (join together).

The fertilised egg is called a **zygote**.
It will start to divide and begin the process of making a new individual.
Sexual reproduction involves two different sets of genes, one from each parent.

The offspring of sexual reproduction show much more variation than those of asexual reproduction.

Where are gametes made?

Gametes are produced in the sex organs.

The male sex organs are called the **testes**.

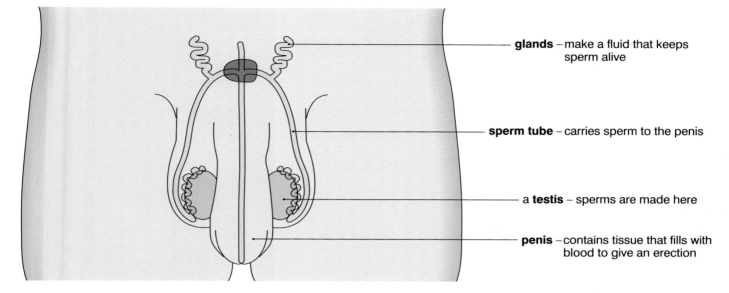

glands – make a fluid that keeps sperm alive

sperm tube – carries sperm to the penis

a **testis** – sperms are made here

penis – contains tissue that fills with blood to give an erection

The female sex organs are called the **ovaries**.

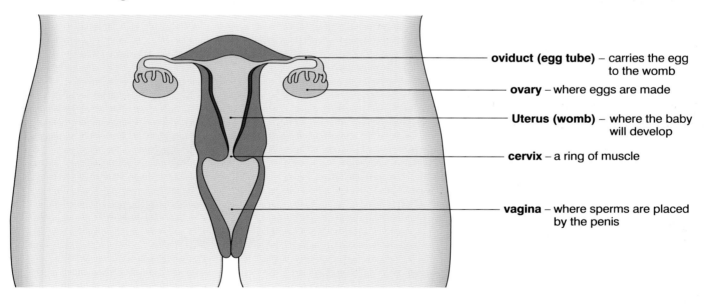

oviduct (egg tube) – carries the egg to the womb

ovary – where eggs are made

Uterus (womb) – where the baby will develop

cervix – a ring of muscle

vagina – where sperms are placed by the penis

Remind yourself!

1 Copy and complete:

Sexual reproduction involves the joining together of ......s. This is called ...... Sperm cells carry the ...... chromosomes and have a ...... that helps them to swim. The female gamete has a large store of ......

2 Copy and complete this table to show the differences between a sperm and an egg.

Sperm	Egg
many produced	
	have a food store
swim using a tail	
	large

Fertilisation in humans takes place inside the female's body.

In some animals, fertilisation takes place outside the female's body – usually in water.

This is called **external fertilisation**.

> **a)** Name an animal that reproduces like this.

Fertilisation is not a very reliable process.
That is why although egg cells are made in small numbers, sperms are made by the million!

In humans, sperm cells are placed inside the woman's body during **sexual intercourse**.

- When a man becomes sexually excited blood flows into the penis.
- This makes the penis become hard and erect, so that it can be placed inside the woman's vagina.
- The man then slides his penis into and out of the vagina.
- This action causes sperm cells in a fluid called **semen** to be released into the vagina.
- The release of the semen is called **ejaculation**.
- Ejaculation is usually accompanied by a pleasant sensation known as an **orgasm**.

The sperm cells now swim through the cervix and into the uterus.
Healthy sperm cells swim at about 5 mm per minute – very fast for their size!
The inside of the woman's body is a very hostile environment for sperm cells.

> **b)** Why are millions of sperm cells released?
> **c)** Work out how far sperm cells could swim in one hour?

Frogs use external fertilisation.

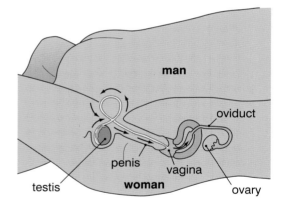

Sexual intercourse.

Fertilisation usually happens in the **oviduct**.

- The egg will have been released from the ovary.
- Eggs cannot swim, but they are moved along by tiny hairs in the oviduct. These hairs waft backwards and forwards like the hairs in our breathing tubes.
- When the egg meets the sperm cells, one of them will get inside.
- Its tail breaks off and its nucleus joins up with the nucleus of the egg.
- This is the moment of fertilisation.
- No more sperm can now enter the egg.
- It starts to divide and very soon will be a ball of cells called an **embryo**.

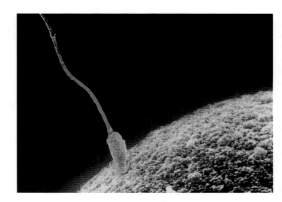

The moment of fertilisation.

d) How many chromosomes will each cell of the embryo contain?

This ball of cells will take 9 months to fully develop. During this time it will be inside the mother's womb.
Here it will receive food and oxygen and be protected from harm.

- In some animals the young do not develop inside the mother's womb.
- Reptiles and birds lay eggs.
- Eggs contain a food store called the yolk. This supplies the developing animal with all the food it needs.
- The eggs are kept warm either by being buried or by the parents sitting on them.

A turtle laying eggs in the sand for protection and warmth.

e) Apart from keeping them warm why else would the turtle eggs be buried?

Remind yourself!

1 Copy and complete:

Sometimes sperm ...... eggs outside the females body. This is called ...... fertilisation. In humans sperm are placed inside the females body by ...... intercourse. The sperm then swim up to the ...... where they might meet an ......

2 a) What causes the penis to become hard and erect?

b) Why does this need to happen?

3 Find out how

i) identical, and
ii) non-identical twins are formed.

- Fertilisation can only take place if a sperm cell meets an egg cell.
- In humans this can only happen on a few days every month.
- This is due to a sequence of events in the female called the **menstrual cycle**.
- Menstrual cycles start to happen when a girl reaches **puberty**.
- This is the time when girls and boys become sexually mature.
- In girls it often begins around the age of 12.
- The cycles will then continue until a woman reaches the age of about 50.
  This is a time called the **menopause**.

The diagram below shows the main events in the menstrual cycle.

This girl is just beginning puberty.

This woman has reached the menopause.

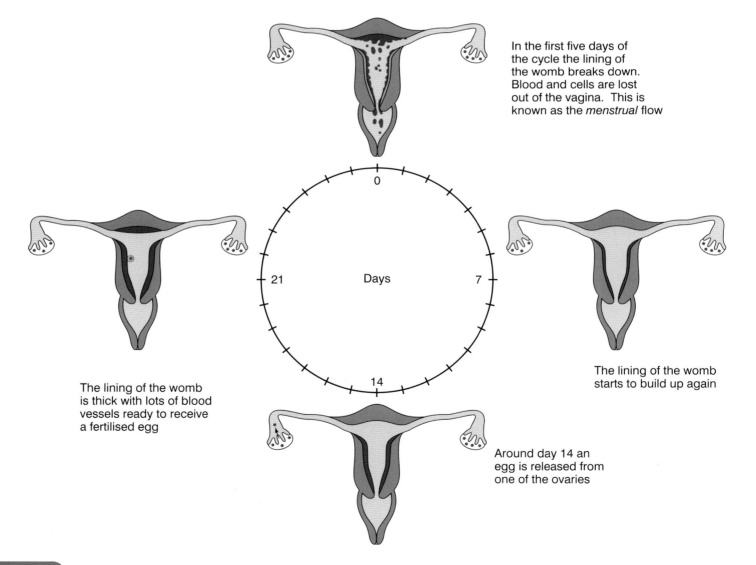

In the first five days of the cycle the lining of the womb breaks down. Blood and cells are lost out of the vagina. This is known as the *menstrual* flow

The lining of the womb is thick with lots of blood vessels ready to receive a fertilised egg

The lining of the womb starts to build up again

Around day 14 an egg is released from one of the ovaries

Controlling the menstrual cycle

The menstrual cycle is controlled by a number of hormones.

> **a)** What are hormones?

These hormones are released from two **glands**.
One is called the **pituitary gland** and it is found at the base of the brain.

> **b)** How will hormones get from the pituitary gland to the reproductive organs?

The other glands are the two **ovaries**.

Once a month sex hormones carry out the following jobs:
- they stimulate an egg in one of the ovaries to become mature
- they stimulate the lining of the womb to become thick again after a period
- they cause an egg to be released around the middle of the month.

If an egg becomes fertilised then the cycle stops until the baby is born.
This happens because the hormones that ripen and release the egg are no longer produced.
The lining of the womb stays thick so that it can support the developing baby.

If the egg doesn't become fertilised then the hormones cause the cycle to start all over again.

> **c)** Why does the menstrual cycle stop once a woman becomes pregnant?

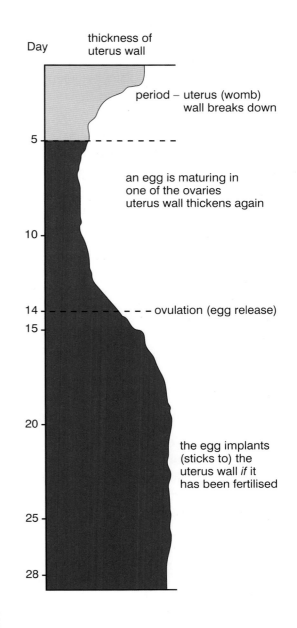

Day — thickness of uterus wall

period – uterus (womb) wall breaks down

5

an egg is maturing in one of the ovaries
uterus wall thickens again

10

14 — ovulation (egg release)
15

20

the egg implants (sticks to) the uterus wall *if* it has been fertilised

25

28

Remind yourself!

1 Copy and complete:

At the start of the menstrual cycle the lining of the …… is lost through the …… About two weeks later an …… is released from one of the ovaries. The lining of the womb now becomes …… again. If the egg is not …… then this lining will …… down, and the cycle begins all over again.

2 How long does the menstrual cycle usually take to complete?

3 Where are the hormones that control the cycle produced?

4 What is a 'period'?

▶▶▶ 16e Controlling fertility

Some women do not produce an egg every month.
This means they have difficulty getting pregnant.
This condition is called **infertility**.

The same hormones that control the menstrual
cycle can be used to treat infertility.
Doctors can give an injection of a hormone called
follicle-stimulating hormone (FSH).
This is one of the hormones from the pituitary gland.
Its job is to make the egg cells in the ovary ripen.
This is known as **fertility treatment**.

Special care must be taken with this treatment.
If lots of eggs ripen at the same time a woman may
find she is expecting four, five or even six children!

Infertility can also be caused by a blocked oviduct.

> **a)** How could a blocked oviduct stop a woman
> getting pregnant?

This type of infertility cannot be treated with hormones.

Another form of fertility treatment is called
in vitro **fertilisation**.
This involves taking mature eggs from the ovary.
They are then fertilised in a laboratory and
returned to the woman's womb.

> **b)** What is the common name for a baby that
> results from *in vitro* fertilisation?

Hormones can also be used to prevent pregnancy.
Oral contraceptives often contain a hormone called
oestrogen.
This hormone stops the pituitary gland from making
FSH.
This means that no eggs are ripened in the ovary.
If a woman does not produce ripe eggs she cannot get
pregnant.

> **c)** What is the common name
> for oral contraceptives?

The Walton family – the result of fertility treatment.

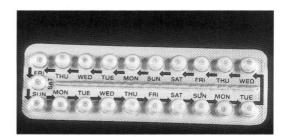

The 'pill' contains hormones.

Summary

The job of reproduction is to make new living organisms.

There are two kinds of reproduction.
Asexual reproduction involves only one parent and no joining of **gametes** (sex cells).
It produces clones (offspring that have the same genes as the parent).
Sexual reproduction involves the joining of two different gametes.
It produces offspring with lots of variation.

In humans the male gametes are the **sperm**, made in the **testes**.
The female gametes are the **eggs**, made in the **ovaries**.
The joining of gametes is called **fertilisation**.

Every month changes happen in a woman's body.
These changes are called the menstrual cycle.
A ripe egg is released from one of the ovaries.
This egg might be fertilised by a sperm.
If it is, it will remain in the womb and develop into a baby.
If it is not fertilised, it will be lost from the vagina along with the lining of the womb.
This loss is called the **menstrual flow**.
The events in the menstrual cycle are controlled by **hormones**.

These hormones can also be used to increase fertility or to prevent pregnancy.

Questions

1 Copy and complete:

   …… reproduction involves only …… parent. It produces offspring with the …… genes as the parent. These offspring are called …… Sexual reproduction involves the joining of two different …… This is called …… This time the offspring show a lot of ……

2 a) What is meant by external fertilisation?

   b) Give two examples of animals that carry out external fertilisation.

3 Using the word 'genes' in your answer, explain why sexual reproduction produces offspring with lots of variation.

4 Give two reasons why a woman can be infertile.

5 What problem can result from fertility treatment with hormones?

6 Find out about the Walton sextuplets.

7 a) Give three examples of plants that reproduce asexually.

   b) Give three examples of other living things that reproduce asexually.

8 Find out about the changes that happen to both boys and girls during puberty.

CHAPTER 17

Selective breeding

▶▶▶ 17a Artificial selection

Look at this photograph of a pedigree bulldog.

a) What is its most obvious feature?

Dogs like these have been bred over many years
to produce their characteristic appearance.
Other dogs are bred for particular abilities:
- the bloodhound is bred for its tracking ability
- the German shepherd for its speed and agility.

These are all examples of artificial selection
(**selective breeding**).

> Artificial selection is the breeding of plants
> or animals to create characteristics useful to humans.

How are the selected features developed?
- We allow animals (or plants) with the features
  we want, to breed.
- Then we let the best of their offspring
  breed with each other.
- This process is repeated many times over.

There are lots of other examples of selective breeding.
- Hereford cattle are bred for the quality of their meat.
- Jersey cattle are bred for their creamy milk.
- Race horses and racing pigeons are bred for
  speed and stamina.

These Herefords produce good meat.

b) What is stamina?

- We breed merino sheep for wool
  but Down sheep are bred for meat.

c) What features might a farmer want in his chickens?

We also carry out selective breeding on plants.
Most plants we eat for food have been
selectively bred.

- Fruit plants like strawberrys are bred to
  produce large juicy fruit.
- Some crop plants like wheat are bred
  to produce large numbers of seeds.
- We breed some plants to have
  a resistance to disease.

Fruit farms are full of selectively bred plants.

The disadvantages of artificial selection

Artificial selection reduces the amount of
variety in a population.
This is because it involves inbreeding.

> **Inbreeding** involves breeding from closely
> related animals or plants.

This means that the offspring do not have
many different alleles.
This may not be a problem now,
but in the future we might need certain features
to cope with changing conditions.
Unfortunately, the alleles that control these features
may well have disappeared.

d) If the climate becomes a lot colder, what feature
might we want in all our farm animals?

Animals like dogs are often bred for 'show'.
Unfortunately the features that breeders want
are not always good for the dogs.

- Short nosed bulldogs have breathing difficulties.
- Basset hounds have long backs that become weak.

The Basset hound has a long back.

Remind yourself!

1 Copy and complete:

Breeding organisms with useful characteristics is
called ...... selection. It is used a lot in growing
plants and animals for ...... Artificial selection
reduces the number of ...... in the population.
This means that some useful ...... might
disappear.

2 What features are bred for in animals used in
sport?

3 Why does inbreeding sometimes cause
problems in pedigree dogs?

▶▶▶ 17b Cloning

In Chapter 16 you read about **asexual reproduction**.
The offspring of this type of reproduction
are all **clones**.

> Clones are organisms that are genetically identical.
> In other words, they have identical genes.

In the plant world there are lots of examples
of natural clones:
- all the potatoes from one type of plant
- strawberry plants that develop from
  one parent plant
- crocus flowers from underground stems.

a) What is the advantage for a plant having
an underground storage organ?

Taking **plant cuttings** is one way of creating clones.
As the cuttings all come from the same plant
they must have the same genes.
To take a plant cutting you must:
- find a shoot on the parent plant
- cut it off just below the point where a leaf joins
- dip the cut end in some rooting powder
- plant the cutting in some compost
- cover the plant pot to keep the atmosphere
  damp.

b) Why must the new shoot be kept damp early on?

c) Why do we dip the shoot in rooting powder?
(Hint: look back at Chapter 6.)

We can take a number of cuttings from one plant.
It is an easy way of growing more of your favourite plant.

Other parts of plants can be used for cuttings.
For example, the African violet can be grown
from **leaf cuttings**.

As well as being cheap and easy to grow,
we also know what our new plants will look like.

Taking cuttings is a type of cloning.

d) Give 3 advantages of growing a plant
from a cutting.

Cloning can also be a 'high-tech' process.

We can grow lots of identical plants using a technique called **tissue culture**. This involves growing plants using just a few cells from the parent plant.

- The first step is to cut off a small piece of tissue.
- This is sterilised without damaging the cells.
- Next, put the tissue onto the surface of some sterile agar jelly.

air tight lid to keep microbes out

glass bottle lets light in

a new plant can be grown from a small bundle of cells

agar jelly contains everything the plant needs to grow

e) What is agar jelly?

- The tissue then divides and grows. If the agar has the right nutrients and hormones, shoots, leaves and roots will grow.
- This makes what we call an **explant**.

f) Why does this process need sterile conditions?

- Eventually the explants will be big enough to be grown in compost.

This technique has a number of advantages:
- it is quick
- it does not take up much space
- plants can be produced all year round
- it produces lots of plants from one parent
- all the plants are genetically identical
- it is a good way to produce large numbers of plants with desirable features.

Tissue culture on a large scale.

Remind yourself!

1 Copy and complete:

Genetically identical organisms are called ......
Potatoes are an example of ...... clones. Plant ...... are clones produced by man. A modern technique of cloning is called ...... culture.

2 Why don't microbes grow on the agar jelly used in tissue culture?

3 What useful nutrients would be needed in the agar jelly?

Meet Dolly the sheep!
Dolly is famous for being the first cloned
farm animal.
What makes Dolly special is that she
was produced from just one parent.
So just how was she produced?
Dolly is a Finn Dorset sheep.

- Scientists took the nucleus from
  a cell from Dolly's mother.
- They put this nucleus into an unfertilised egg
  from a Scottish Blackface sheep in the lab.
- When this new cell started to divide,
  it was put into the uterus of this sheep.
- 148 days later Dolly was born.
- She was genetically identical to her mother.

Although two sheep were involved scientists only
used the genes from **one** of them.

a) Why might farmers want to clone their animals?

Animal cloning means that we can preserve useful
characteristics.
A cow might produce a large amount of milk.
Cloning this cow would enable a farmer
to produce lots of cows, all giving lots of milk.

Human cloning – a reality?

If animals can be cloned then so can humans.
But in Britain experiments on human embryos
are banned.
People are concerned that this technology could
get into the wrong hands.
Religious and scientific experts think that
trying to copy humans is wrong.
It ought to be possible to create an identical
looking human.
However someone's personality is not just
controlled by genes.
So a cloned human would not be completely
identical to its 'parent'.

b) What is the difference between
human cloning and the work of
the fictional character Dr Frankenstein?

'Dolly' the sheep – a cloned farm animal.

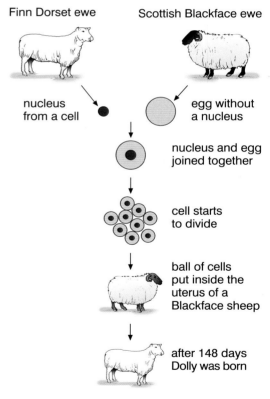

How Dolly was created.

The most common way of cloning animals is called **embryo transplantation**.
People have used this method to produce lots of identical farm animals.

Female sheep and cows often produce just one offspring at a time.
We use embryo cloning to help farmers rapidly increase their numbers of animals.
How does embryo transplantation work?

- Eggs from the best female are mixed with sperm from the best male.
- An egg is fertilised and it starts to divide producing an embryo.
- This embryo is carefully split into lots of identical embryos.
- These embryos are put into separate females called **surrogates**.
- They develop and are born normally.

These new animals will have the useful characteristics of the parents.

They will not be exact copies of either parent but they will be clones of each other.

Breeders can use this method to breed rare or expensive animals.
For example, sperm and eggs can be collected from expensive angora goats.
These goats produce valuable wool.
The embryos produced can be put into ordinary less expensive goats.
These surrogate goats will eventually give birth to angora kids.

> **c)** When newspapers talk about women being surrogate mothers what do they mean?

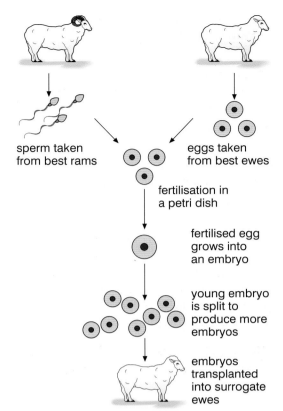

Stages in embryo cloning.

An angora goat

Remind yourself!

1 Copy and complete:

Animals can be cloned by ...... transplantation. An ...... is split into lots of identical cells. Each embryo is then put inside a ...... mother. Here it develops normally until ......

2 Give two reasons why cloning of farm animals is useful.

3 What do you think about human cloning?

Discuss the points for and against with your friends.

In Chapter 8 you read about how your body controls its blood sugar level.
To do this your body makes hormones.
One of these hormones is **insulin**.

a) What disease is caused by a lack of this hormone? (Hint: look in Chapter 8.)

Diabetics need regular injections of insulin.
Until quite recently most of this hormone came from pigs.
This created some problems:

● insulin from another species does not work as well as human insulin
● some people react badly to pig insulin.

Today we make most insulin by **genetic engineering**.

Genetic engineering means transferring genes from one type of cell to another.

Genes carry all the instructions for how your body works.
The technique used is to get cells from **microbes** to make useful chemicals.

● Human genes are cut out of cells using special enzymes.
● They are then inserted into bacterial cells.
● These cells then divide making exact copies of themselves – and the human gene.

b) What do we call exact copies in biology?

Bacteria divide very quickly and very soon there will be thousands, each with the human gene.
Given the right conditions the bacteria will make insulin which can be harvested.
The advantages of this technique are:

● the insulin is identical to human insulin
● it is made in large quantities
● it is cheap to produce.

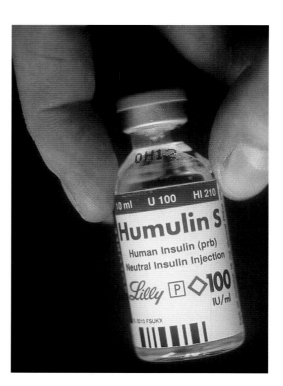

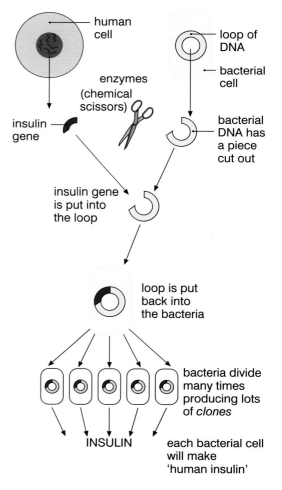

human cell

loop of DNA

bacterial cell

enzymes (chemical scissors)

insulin gene

bacterial DNA has a piece cut out

insulin gene is put into the loop

loop is put back into the bacteria

bacteria divide many times producing lots of *clones*

INSULIN

each bacterial cell will make 'human insulin'

Another interesting example of genetic engineering is making **factor VIII** (eight).
We have factor VIII in our blood.
Without it our blood will not clot when we cut ourselves.
This is the cause of the disease called **haemophilia**.

c) What else is needed for our blood to clot?

Scientists have transferred the gene for factor VIII into fertilised sheep eggs.
These egg cells divide to form an embryo.
Each time the cells divide the gene is copied.
Eventually a new lamb will be born, and each of its cells will contain the gene.
The factor VIII is then released in the sheep's milk.
Animals with new genes like this are called **transgenic animals**.
Human insulin has also been produced in this way.

The future?

The next step for scientists is to try to put new genes into humans.
Why should they want to do this?
Consider the inherited diseases in Chapter 15.
Cystic fibrosis for example is caused by a faulty gene.
Scientists are experimenting with ways to replace this faulty gene with a normal one.
The big problem is getting the new gene into enough of the patient's cells.
At the moment scientists are experimenting with inhalers.
Each inhalation contains millions of copies of the correct gene.
Some of these will enter the cells lining the lung.
This technique is called **gene therapy**.

This sheep will produce factor VIII in its milk.

If gene therapy works treatment for CF might involve a simple inhaler.

Remind yourself!

1 Copy and complete:

Genetic engineering means transferring …… between different cells. It is used to make human …… to treat diabetics. The insulin …… is put into …… These bacteria are then able to make insulin …… to human insulin.

2 Why are bacteria able to produce large quantities of insulin?

3 Where in the body do scientists get the human insulin gene from?

4 If it works why will gene therapy be so important?

▶▶▶ 17e GM foods

Look at the photograph opposite.
It shows members of Greenpeace
vandalising a maize crop in Norfolk.

a) What is Greenpeace?

This maize crop had been grown from
genetically modified seeds.

> **Genetically modified** plants have had their genes
> altered by humans.

Using genetic engineering, scientists have altered
the genes of many plants.
For example they have produced:

- maize resistant to pests
- soya beans that are resistant to weed killers
- vegetarian cheese
- tomatoes that stay fresh for longer.

These (and other) **GM foods** have some important
potential benefits.
For example:

- farmers will need to use fewer chemicals
  on their land
- food production could be increased,
  helping to solve global hunger
- shoppers will be able to buy better quality
  fresh foods all year round.

GM tomatoes stay fresh longer.

So why are organisations like Greenpeace worried?
Two particular concerns are:

- the danger of modified genes transferring to wild plants
  or animals
- they are unhappy that this technology is owned by a small
  number of very big companies.

The general public have also been put off buying GM foods.
This is because of badly researched stories in the media.
These stories talked about 'Frankenstein' foods
and without a lot of evidence implied
that they could be very harmful to humans.

b) Ask your family how they feel about
GM foods.
Do they buy them or avoid them?

Summary

We use **artificial selection** to produce new varieties of organisms.
Individuals with useful characteristics are chosen and used for breeding.
Artificial selection is also known as **selective breeding**.
Selective breeding reduces the number of alleles in a population.
This means that there is less variety and less chance of populations adapting
to environmental changes in the future.

Clones are **genetically identical** organisms.
Clones occur naturally and this is an example of **asexual reproduction**.
Taking cuttings is an example of cloning.
More modern examples include:

- **tissue culture**
- **embryo transplantation**

Genetic engineering involves transferring genes between cells.
Human genes can be put into other organisms.
The human insulin gene has been transferred to bacteria.
These bacteria now make human insulin.
Human genes have also been transferred to other species, e.g. sheep.
This enables the animals to make chemicals useful to humans.

GM foods have had their genes altered.
This produces useful features such as longer life or pest resistance.

Questions

1 Copy and complete:

New varieties of organisms are produced by
...... selection. This is also known as ......
breeding. The organisms with the best ......
are ...... with each other. Selective breeding
reduces the number of ...... in a population.
This means that the population might not be
able to ...... to environmental changes in the
future.

2 Why is it a good idea to keep a few examples
of rare and unusual animals, or seeds of rare
plants?

3 Why are the enzymes used in genetic
engineering often called 'chemical scissors'?

4 Do some research on GM foods.

Write a report or produce a poster
explaining their advantages and
disadvantages.

5 What does the phrase 'bred for show' mean?

How has selective breeding caused problems
for some pedigree dogs?

6 What are the advantages of producing new
plants by tissue culture?

CHAPTER 18

EVOLUTION

▶▶▶ 18a What is evolution?

Evolution is change that happens
over a period of time.
Look at these drawings of bicycles.
One is a 19th-century penny farthing
and the other a 21st-century mountain bike.
Bicycles have evolved because technology
has changed.
There are stronger and lighter materials now.
Brakes and tyres are more advanced.

Many plants and animals have **evolved** to
suit their environmental conditions.
If an animal or plant species does not adapt
to its surroundings it will die out.

> Evolution is the gradual change in the
> characteristics of a species.

What is a species?

The word species is used a lot,
but it is not always well understood.

Two animals are said to belong to
the same species if:

> '… they can breed together and produce
> fertile offspring'

If the offspring are fertile this means that they
could also go on to breed.
The opposite of being fertile is **sterile**.

A good example of this is the **mule**.
Mules are the product of a **donkey**
breeding with a **horse**.
But because the donkey and the horse are
separate species, mules are sterile.
The only way to get more mules is
to breed more donkeys with more horses.

*A mule – the product of a horse and a
donkey.*

The theory of evolution

This theory says that:

> All species of living things that exist today evolved from more simple forms of life.

These simple life forms first appeared 3 billion years ago.

The first living organisms were simple bacteria, plants and animals found in the sea.
Over millions of years these creatures have evolved into much more complicated organisms – such as humans!

Extinction

Between 3 billion years ago and today many plant and animal species have disappeared.
They have become **extinct**.

a) Name a famous group of animals that has become extinct?

Animals that are in danger at the moment – like the panda – are threatened by human activities.
But some creatures have become extinct for other reasons.

What happened to dinosaurs?

Dinosaurs were a kind of reptile.

b) What are the characteristic features of reptiles?

These died out long before humans first appeared.
It is possible that their food supply died out.
Or maybe they were wiped out by the effects of giant meteorites hitting the earth.

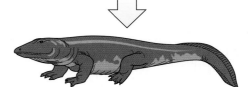

simple creatures existed 3 billion years ago

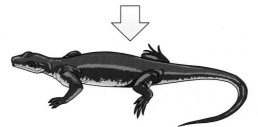

fish first appeared 400 million years ago

amphibians first appeared 300 million years ago

reptiles first appeared 250 million years ago

mammals first appeared 200 million years ago

birds first appeared 150 million years ago

Remind yourself!

1 Copy and complete:

Evolution is a …… change in the characteristics of a …… These changes enable species to …… to a …… environment. A species that cannot adapt will become ……

2 What type of creatures did mammals evolve from?

3 Do some research on dinosaurs. Prepare a presentation to give to the rest of your class.

▶▶▶ 18b Fossils

A lot of what we know about evolution
has come from studying fossils.

> Fossils are the preserved remains of animals
> and plants from millions of years ago.

There are three main ways that fossils are
formed:

- The hard parts of living organisms do not
  decay easily. These are things like bones and shells.
  They get buried underground by being
  covered in fine particles called **sediment**.
  Eventually the sediment turns into rock,
  and the hard parts of the body are
  replaced by tough minerals.
  The result is a perfect stone copy of these parts of the
  creature.
- Sometimes even the soft parts do not
  decay.
  This happens when the conditions are
  not right for microbes. For example,
  when the creature is buried in ice or in
  an acid peat bog.
  Ancient insects are often preserved in
  amber.
  This is a sap from trees that forms a hard
  coating around the insect.
- If the soft parts decay slowly enough
  they can be replaced by minerals.
  Many fossils of ancient plants have
  been formed this way.
  They are often found in coal.

a) Why do animals not decay if they get buried in ice?

It is not just plants and other animals
that can be fossilised, humans can too.
A good example is the famous Lindow man
found in peat near Wilmslow in Cheshire.

The archaeopteryx – an early bird.

Amber sets hard trapping insects.

Why are fossils important?

Fossils can show us how different organisms have changed over time.

Geologists study the Earth and how it was formed.
They can work out how old rocks are.
If they know this, they also know the age of the fossils in these rocks.
Using this information they have been able to work out the **fossil record**.

The fossil record shows the order in which animals and plants evolved.
A good place to study this record is the Grand Canyon in America.
This is a deep gorge with rocks as old as 2 billion years.
Geologists have found many fossils here.
The missing animals are the most recent ones like birds and mammals.

Fossil hunting.

The Grand Canyon.

The appearance of mammals

Humans are examples of mammals.

b) What are the characteristics of mammals?

Mammals first appeared around 70 million years ago.
(Not long after the dinosaurs disappeared in fact.)
Humans are the most recent mammals.
We only appeared about 2.5 million years ago, and so far we have proved to be very successful animals.

c) What do you think is the main threat to human survival on Earth?

Remind yourself!

1 Copy and complete:

Fossils are the …… remains of ancient plants and animals. They show us how different organisms have …… over time. The …… record shows us which organisms lived at …… time.

2 Which parts of animals and plants decay quickly?

3 How do these parts get turned into fossils?

4 Why do geologists need to know the age of rocks?

Cod is a very familiar fish.
Female cod can lay millions of eggs.
But we hear on the news about how
North Sea cod stocks have almost run out.
What has gone wrong?
Why aren't we overrun with cod?

In reality many eggs do not become adult
fish.
They are often eaten.
Many of those that do become adults
end up as part of a fish supper!

> **a)** Frogs lay thousands of eggs, so why
> aren't we overrun with frogs?

Many animals and plants produce lots of
offspring.
But a large number of these do not survive.

> **b)** Apart from being eaten, why else do many
> animals not survive?

> The individuals that survive are those
> **best adapted** to their environment.

These individuals survive long enough to breed.
Then they pass on the alleles for their 'survival
features' to their offspring.
These offspring now also stand a good chance
of surviving.

Gradually over many generations there will be
more and more individuals who are well adapted.

> **c)** What feature(s) might help a cod to survive?

The idea of only the best adapted organisms
surviving is known as 'the survival of the fittest'.

*Most fish lay large numbers of eggs.
Here eggs are being collected from a fish in
a process called stripping.*

*The kingfisher is well adapted to feed from
rivers.*

This is more correctly known as **natural selection**.

The theory of natural selection was put forward by **Charles Darwin**.
Darwin was a British scientist who travelled the world studying wildlife.
He based his theory on these observations:

- organisms produce lots of offspring
- there is always variation between members of a species
- there is always a struggle for survival between members of a species
- individuals with useful characteristics are more likely to survive
- those that survive will breed
- the next generation will have better survival characteristics.

Darwin used this idea to explain the changes that happen in evolution.

Charles Darwin.

Natural selection and giraffes

Have you ever wondered how giraffes got such long legs and necks?

- In the original population there would have been *some* very tall giraffes.
- These individuals would have had an advantage over shorter giraffes.
  They could feed more easily from leaves on trees.
- Therefore they would have survived better than their smaller cousins.
- These giraffes would have bred and passed the genes for long necks onto their offspring.
- Over a long period of time more and more giraffes were born with long legs and necks.

Sorry mate you'll have to evolve a bit more

Remind yourself!

1 Copy and complete:

   Those organisms ...... adapted to their ...... are more likely to survive. Organisms that survive are more likely to ...... and pass on their ...... to their offspring. This is called the theory of ....

2 Why do you think Darwin did not mention genes when he wrote about his theory?

3 Antelopes in Africa have good camouflage.

   Why do you not see antelopes with bright, coloured stripes?

Doctors often prescribe drugs called antibiotics.
They are used to kill bacteria.

a) Which was the first antibiotic?

Today there is a lot of concern about overuse
of antibiotics.
In a population of bacteria there might be
one or two **resistant** individuals.
This means that the antibiotic will not
kill them.

These survivors will quickly reproduce
and pass on their resistance.
Very soon there will be millions of bacteria,
all resistant to the drug.

The more antibiotics we use, the more likely
it is that resistance will develop.
As a result:
- antibiotics should only be used if
  absolutely necessary
- patients should always finish a course
  of tablets
- farmers should use fewer antibiotics
  when raising their animals.

Resistance doesn't only affect antibiotics.
In the 1950s, a rat poison called **warfarin**
was developed.
At first it was very effective and killed a lot
of rats.
But some rats were resistant to warfarin.
These rats passed on their resistance
to their offspring.
Today warfarin is less effective and even
large doses will not kill some rats.

Rats present a health risk.

b) Why are rats considered to be pests?
c) Why is it becoming more difficult to
  get rid of rats?

Summary

Evolution is the gradual change in the characteristics of a species.
All living things that exist today evolved from simpler organisms.
These first simple organisms developed about 3 billion years ago.

Some organisms have become extinct.
Extinction happens when some part of an organism's environment changes, and the organism is unable to adapt to it.

Evidence for evolution comes from the **fossil record**.
Fossils are the preserved remains of ancient plants and animals.
The hard parts of organisms can be preserved in rock.
Sometimes the soft tissues do not decay because they are buried in ice or in acid peat bogs.
Small insects have been found preserved in solidified tree sap called amber.
Geologists work out the age of rocks and can therefore work out how old fossils are.
Fossils show the changes that have occurred in living organisms over millions of years.

Charles Darwin explained evolution using his theory of **natural selection**.
Organisms that are adapted to their environment are more likely to survive.
These organisms will breed and pass on their useful characteristics in their genes.

Questions

1 Copy and complete:

A ...... change in the ...... of a species is called ...... If a species does not evolve it will become ...... The theory of ...... selection helps to explain evolution.

Organisms that are adapted to their ...... will survive and ...... They will pass on their useful ...... to their ...... This theory was developed by .... He called it the survival of the ......

2 What were the observations that Darwin based his theory of natural selection on?

3 The dodo is a famous example of an extinct animal.

Find out about the dodo and write a report on it.

4 Typhoid is a disease usually controlled by an antibiotic.

In 1972 14 000 people in Mexico died of typhoid because the usual antibiotic did not work.

Explain why the antibiotic did not work. (Hint: re-read the last page of this chapter.)

5 Do some research and write a report about the reaction of the public when Darwin first published his theory of natural selection.

▶ Variation

1 The features of humans are either inherited or are caused by environmental influences.

**accidental loss of an arm   blood group
eye colour   gender/sex   mass**

(a) From the features listed choose:

(i) one which is controlled by environmental influences only   (1)

(ii) one which is controlled by inheritance only   (1)

(iii) one which is controlled by both environmental influence and inheritance. (1)

(b) Which part of the cell nucleus enables features to be passed from one generation to the next?   (1)

(c) What is meant by the term **dominant allele**?   (1)

(OCR)

2 A scientist wants to find the genes which give a plant its flower colour.
Choose a word from each list to complete the sentences that follow.

(a) **cold   damp   dry   hot**
First, a lot of the plants are grown from cuttings. To make their roots develop they must be grown in an atmosphere which is ....

(b) **cell wall   cytoplasm   membrane
nucleus**
Then the scientist removes chromosomes from leaf cells.
Chromosomes are found in the ....

(c) **body cells   chloroplasts   gametes
sex cells**
The chromosomes are in pairs.
This means that leaf cells are ....

(d) **alleles   clones   recessives   variations**
Different flower colours are caused by different forms of the same gene.
Different forms of the same gene are called
....   (4)

(AQA 2001)

3 The diagrams show four ways in which human twins may be formed.

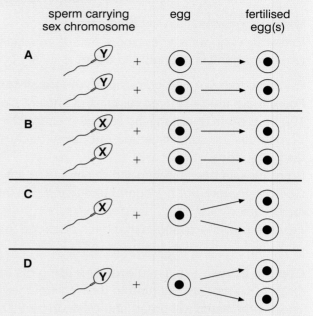

Which diagram, **A**, **B**, **C** or **D**, shows the process which will produce genetically identical twin boys? Explain the reason for your choice.   (3)

(AQA 2001)

4 (a) Use words from the list to complete the sentences.

**alleles   chromosomes   gametes
genes   mutations**

The nucleus of a cell contains thread-like structures called ....
The characteristics of a person are controlled by ...... which may exist in different forms called ....   (3)

(b) The drawing shows some of the stages of reproduction in horses.

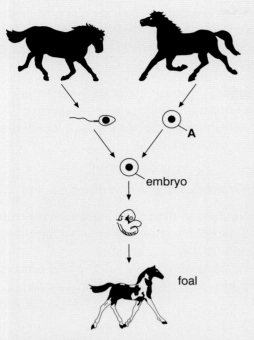

embryo

foal

(i) Name this type of reproduction.   (1)

(ii) Name the type of cell labelled **A**.   (1)

(c) When the foal grows up it will look similar to its parents but it will **not** be identical to either parent.

(i) Explain why it will look similar to its parents.   (1)

(ii) Explain why it will **not** be identical to either of its parents.   (2)

(AQA (NEAB) 1999)

5 A gardener took four cuttings from the same plant and put them in compost. He kept them in different conditions. The diagrams show each cutting two weeks later.

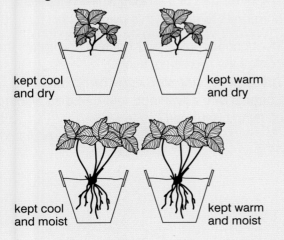

kept cool and dry

kept warm and dry

kept cool and moist

kept warm and moist

(a) Use information from the diagrams to answer this part.

(i) The most important condition needed for cuttings to develop is that they should be kept ....   (1)

(ii) Explain why you chose this condition.   (2)

(b) Gardeners often grow new plants from cuttings instead of from seeds. Give a reason for this.   (1)

(AQA (NEAB) 2000)

6 This drawing shows the chromosomes from a human cell.

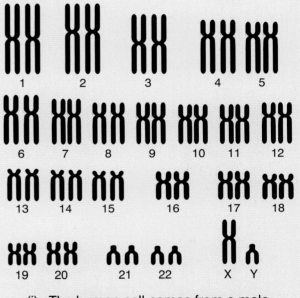

(i) The human cell comes from a male. How can you tell that the chromosomes are from a male?   (1)

(ii) How can you tell that the chromosomes do **not** come from a sex cell?   (1)

(AQA (NEAB) 2000)

▶ Inheritance

7 Read the following and answer the questions which follow:

Cystic fibrosis is a genetic disease which affects the pancreas. The ducts of the pancreas become blocked. The disease also affects the mucus-producing glands of the bronchioles and these produce very large amounts of thick sticky mucus. This makes breathing difficult and reduces the ability of the respiratory system to remove microbes.

The disease cannot be cured but some of the symptoms can be eased by the use of antibiotics, physiotherapy and careful control of the diet.

(a) What is a genetic disease? (1)

(b) Explain why cystic fibrosis sufferers find breathing difficult. (1)

(c) Suggest why antibiotics can help sufferers of cystic fibrosis. (1)

(d) Tracy and Peter do not suffer from cystic fibrosis but one of their children, David, does suffer from it. Here is the family tree:

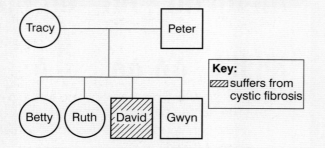

Use the following letter: **N** = normal (dominant) and **n** = cystic fibrosis (recessive).
What are the genotypes of:

(i) Tracy (1)

(ii) Peter (1)

(iii) David (1)

(iv) What are the **two** possibilities of Gwyn's genotype? (2)

(e) Name any other genetic disease. (1)

(WJEC)

8 People inherit their sex. They can also inherit certain diseases from their parents.
Choose words from each list to complete the sentences that follow.

(a) eggs   genes   nuclei   sperms
In your body cells, one of the pairs of chromosomes carries the ...... which make you male or female.

(b) XY   XX   YY
In females, the sex chromosomes are ....

(c) circulatory   digestive   nervous   reproductive
Huntington's chorea is an inherited disease. It affects the ...... system.
The parts of a male's body which produce sperm cells belong to his ...... system. (4)

(AQA (NEAB) 2000)

▶ Reproduction

9 This question is about the monthly menstrual cycle.
During the cycle the thickness of the womb (uterus) lining changes and an egg is released from an ovary. The cycle is controlled by hormones.

(a) Which part of the female reproductive system releases one of these hormones? (1)

(b) Choose a word from this list to complete the sentence that follows.

**liver   pancreas   pituitary gland
salivary gland**

The other part of the body which releases hormones to control this cycle is the .... (1)

(c) Most contraceptive pills contain hormones that control the menstrual cycle.
These hormones stop eggs being released.
Suggest **one** advantage and **one** disadvantage of using hormones in this way. (2)

(AQA 2001)

10 In women, two hormones control ovulation (the release of eggs from the ovaries).

(a) (i) What is a hormone? (1)

(ii) How are hormones transported around the body? (1)

(b) Hormones can be used to control human fertility. Describe the benefits and problems that might arise from using hormones in this way. (4)

(AQA (NEAB) 1998)

▶ **Selective breeding**

11 Many years ago nearly all wheat plants had very long stalks. Then big improvements were made in the number and size of seed grains.

• Long stalks made the tops of the plants heavy and caused them to bend over.

• To solve this problem, scientists used artificial selection to develop plants with short stalks.

(a) Put the sentences **A–D** below into the correct order so that they tell you how to develop short-stalked wheat plants.

A Repeat this for many generations.

B Select plants with shorter stalks.

C Grow the seeds into new plants.

D Breed these together to produce seeds.

(4)

(b) Suggest a reason why farmers wanted wheat plants with short stalks in our British climate.

(1)

(AQA (NEAB) 1999)

12 Breeding guide dogs for the blind is an example of artificial selection.

When guide dog training started:

• people chose dogs which were alert and able to think for themselves;

• these dogs were trained and then tested;

• result: 2/10 dogs passed.

When it's done now:

• Successfully trained guide dogs are used for breeding.

• Their puppies are trained and then tested;

• Result: 9/10 dogs pass.

The sentences **A–D** are in the wrong order. Write them again in a flow diagram like this so that they show how selection was used to develop modern guide dogs.

A Dogs passing the tests used for breeding.

B Alert, quick thinking dogs chosen for training.

C 9 out of 10 puppies successfully trained.

D Dogs given hard tests. (4)

↓

↓

2/10 dogs passed the tests

↓

↓

(4)

(AQA 2001)

13 The drawings show a wild pig and an English Large White pig. The drawings are to the same scale.

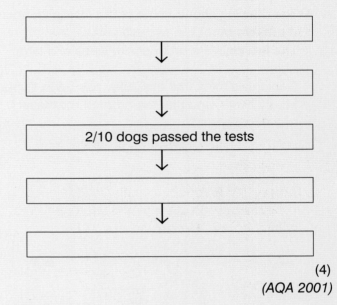

wild pig

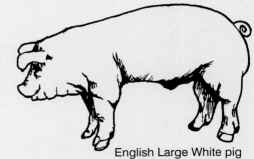

English Large White pig

(a) The English Large White pig has been produced from the wild pig by selective breeding.
Use information from the drawings to give **two** characteristics which farmers selected when breeding the English Large White pig. (2)

(b) Give **two** advantages of producing new breeds of animals. (2)

(AQA (NEAB) 1998)

▶ Evolution

14 The drawing shows some of the fossils found in the layers of rock in two cliffs.
The two cliffs are on opposite sides of a large valley.
Geologists think that the valley has been carved out by rivers, and that the order of rock layers has not changed.

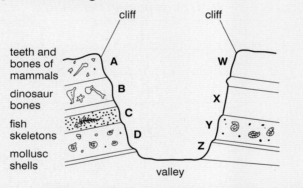

(a) (i) Which of the rock layers, **A**, **B**, **C** or **D**, is the oldest? (1)

(ii) Give the letters of **two** layers of rock on opposite sides of the valley that are the same age. (1)

(b) How do fossils provide evidence for the theory of evolution? (2)

(AQA (NEAB) 1998)

15 The diagram shows two types of peppered moth found on the trunks of trees in both city and countryside areas.

dark moth   light moth

Read the following passage about the peppered moth.

> The peppered moth is often used to provide evidence for evolution. Before the Industrial Revolution the moths found were light in colour. They were harder for birds to see on light coloured tree trunks. Fewer moths are eaten by birds if they are well hidden.
> During the Industrial Revolution, many trees in large cities became black with soot. In the daytime the moths rest on tree trunks. Dark coloured moths were better hidden on the trees than light coloured ones.
> Records showed that few light coloured moths survived in cities and few dark coloured moths survived in countryside areas. In 1956 the Clean Air Act gradually reduced the amount of smoke pollution. After 1956 the number of light coloured moths in the city areas gradually increased.

(a) Use this information to answer the following questions.

(i) In which area were the greatest number of light coloured moths found after the Industrial Revolution? (1)

(ii) Describe and explain the effect of the Clean Air Act on the numbers of the two types of moth in city areas. (4)

(iii) Name the process of evolution described in the passage. (1)

(OCR)

GCSE exams have 20% of the marks awarded for coursework.
Your teacher has to assess your practical skills.
You are given marks in 4 areas:

P   **Planning** to collect evidence.

O   **Obtaining** the evidence.

A   **Analysing** your evidence and drawing conclusions.

E   **Evaluating** your evidence.

Your teacher will mark you against checklists of points
to look out for in your work. You can see these for yourself
in the sections that follow.
Try to cover all the points, if they apply to your task, working
from 2 marks upwards.

P   Planning

Choosing apparatus

It is important to use the most suitable equipment.
For example, if you are measuring 50 cm^3 of water,
you should use a measuring cylinder. It is not a good idea
to just use a beaker with a 50 cm^3 mark on it. Why not?

Deciding on how many readings

You will need to think about how many measurements
or observations to make in your experiments.
If you plan to show your results on a line graph,
aim to collect 5 different measurements.
And if the measurements are tricky to make,
you should repeat them. Taking the average of
the measurements will make them more **reliable**.

Safety

Make sure your plan is a safe one.
You must check to see if the chemicals you plan to use are
hazardous. Will you be using sharp knives to cut plant tissue?

Practical skills are important!

Checklist for skill P PLANNING YOUR WORK	
Candidates:	**Marks awarded**
• plan a simple method to collect evidence	2
• plan to collect evidence that will answer your questions • plan to use suitable equipment or other ways to get evidence	4
• use scientific knowledge to: – plan and present your method – identify key factors to vary or control – make a prediction if possible • decide on a suitable number and range of readings (or observations) to collect	6
• use detailed scientific knowledge to: – plan and present your strategy (the approach you have decided on) – aim for precise and reliable evidence – justify your prediction if you made one • use information from other sources, or from preliminary work in your plan	8

O Obtaining your evidence

Making accurate measurements and observations

Accuracy is important, as well as taking care in
checking your results.

Common mistakes include:
- not checking your balance is on zero when measuring mass,
- spilling powders before, during or after finding their mass,
- not reading to the **bottom** of the meniscus (curve)
  when measuring volumes of liquid.

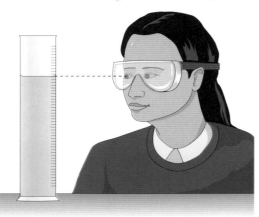

Checklist for skill O OBTAINING YOUR EVIDENCE	
Candidates:	**Marks awarded**
• use simple equipment safely to collect some results	2
• make adequate observations or measurements to answer your questions • record the results	4

• make observations or measurements,
  – with sufficient readings,
  – which are accurate, and
  – repeat or check them if necessary
• record the results clearly and accurately
	6

• carry out your practical work
  – with precision and skill,
  – to obtain and record reliable evidence,
  – with a good number and range of readings
	8

You should consider if using **data-loggers** will improve
the quality of the evidence you collect.

If one of your results seems unusual, make sure you
repeat it. If you find that it was an error, you don't
have to include it in your final results.
(But do comment on it in your evaluation!)

Recording your results

You will often record your results in a table.
In the first column of your table you put the thing (variable)
that you changed in your experiment.
In the second column you put the thing (variable) that you judged or
measured.
Look at the examples below:

Types of exercise	Heart rate during exercise
Sitting	60
Step ups	75
Jog on spot	85
Star jumps	105
Sprinting	130

Relative light intensity	Volume of gas given off (cm^3)
5	1.5
10	2.5
15	3.0
20	3.5
25	4.0

↑ *You decided on the exercise.*   ↑ *You measured your heart rate.*

↑ *You changed the intensity of the light.*   ↑ *You measured how much gas was given off in five minutes.*

A   Analysing **and drawing conclusions**

Drawing graphs and bar charts

Once you have recorded your results in a table,
drawing a graph will show you any patterns.

Whether you draw a bar chart or a line graph
depends on your investigation.
Here's a quick way to decide which to draw:

> If the thing (variable) you change is described in words,
> then draw a **bar chart**.
> If the thing (variable) you change is measured,
> then draw a **line graph**.

Let's look at the results from the tables on the previous page:

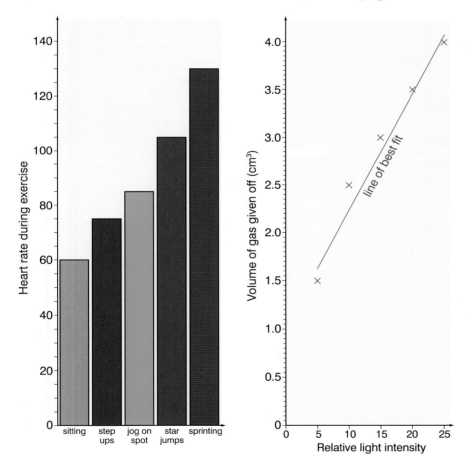

Notice that the thing (variable) you change always goes along the
bottom of your graph. The thing (variable) you measure
goes up the side.

You can then see if there is a link between the two variables.
For example, the greater the light intensity,
the more gas was given off. See the line graph above.

Then try to explain any patterns you spot on your graphs
using the ideas you have learned about in science.

E Evaluating

When you have drawn your conclusions, you should now think about how well you did your investigation.

Ask yourself these questions to see if you could have improved your investigation:

- Were my results accurate?

- Did any seem 'strange' compared to the others? These are called **anomalous** results.

- Should I have repeated some tests to get more reliable results? Could I improve the method I used?

- Did I get a suitable range of results? The range is the spread of values you chose.

  For example:
  if you were seeing how temperature affected something, choosing to do tests at 20°C, 21°C and 22°C would not be a good range to choose!

- If there is a pattern in my results, is it only true for the range of values that I chose? Would the pattern continue beyond this range?

- Would it be useful to check your graph by taking readings *between* points?

  For example:
  if you have a sudden change between two points, why not do another test to get a point half way between? This would check the shape of the line you get.

- How would I have to change my investigation to get the answers to the questions above?

Checklist for skill E EVALUATING YOUR EVIDENCE	
Candidates:	**Marks awarded**
• make a relevant comment about the method used or the results obtained	2
• comment on the accuracy of the results, pointing out any anomalous ones • comment on whether the method was a good one and suggest changes to improve it	4

• look at the evidence and:
   – comment on its reliability,
   – explain any anomalous results,
   – explain whether you have enough to support a firm conclusion
 • describe, in detail, further work that would give more evidence for the conclusion
	6

As you study Science, you will need to use some general skills along the way.
These general learning skills are very important, whatever subjects you take or job you go on to do.

The Government has recognised just how important the skills are by introducing a new qualification.
It is called the **Key Skills Qualification**.
There are 6 key skills:

- **Communication**
- **Application of number**
- **Information Technology (IT)**
- **Working with others**
- **Problem solving**
- **Improving your own learning**

The first 3 of these key skills will be assessed by exams and by evidence put together in a portfolio.
You can see what you have to do to get the first level in the sections below.

Communication

In this key skill you will be expected to:

- **Hold discussions**
- **Give presentations**
- **Read and summarise information**
- **Write documents**

You will do these as you go through your course and producing your coursework will help.

Look at the criteria below:

What you must do ...
Take part in discussions.
Read and obtain information.
Write different types of document.

Application of number

In this key skill you will be expected to:

- **Obtain and interpret information**
- **Carry out calculations**
- **Interpret and present the results of calculations**

What you must do ...
Interpret information from different sources.
Carry out calculations.
Interpret results and present your findings.

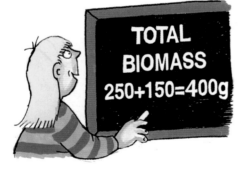

Information Technology

In this key skill you will be expected to:

- **Use the internet and CD ROMs to collect information**
- **Use IT to produce documents to best effect**

What you must do ...
Find, explore and develop information.
Present information, including text, numbers and images.

▶▶▶ Revising and doing your exams

When you walk into your Science exam, you will already
have your coursework marks completed.
If you do Modular Science, you will also have your test marks.
But your final exam is still the biggest part of your GCSE.
So it's important that you prepare well and feel good on the day.

Plan your revision in the weeks leading up to the exams.
Don't leave it too late!

The question 1's at the end of each chapter are a good way
to revise the Summaries. These contain the essential notes
you need.
Then try the past paper questions (coloured pages) on the chapter.
If you get stuck, ask a friend or your teacher the next day for help.

Just sitting there (especially in front of the TV!), reading your notes
isn't good enough for most people. *Active* revision is better.
And don't try to revise for too long without a break.
Do 25 minutes, then promise yourself a 10 minute rest.
This works better than trying to revise non-stop.

Work out your best way of learning!

So you've finished your revision (it's too late to worry about that
anyway!), and it's the day of the exam. What will you need?
Remember to bring:

- Two pens (in case one runs out).
- A pencil for drawing diagrams.
- An eraser and ruler.
- A watch for pacing yourself during the exam.
  (It might be tricky to see the clock in the exam room.)
- A calculator (with good batteries).

You will feel better if you know exactly what to expect.
So collect all the information about your exam papers.
You can use a table like the one shown below:

Date, time and room	Subject, paper number and tier	Length (hours)	Types of question: – structured? – single word answers? – longer answers? – essays?	Sections?	Details of choice (if any)	Approximate time per page (minutes)
4th June 9.30 hall	Science (Double Award) Paper 1 (Biology) Foundation Tier	$1\frac{1}{2}$	Structured questions (with single-word answers and longer answers)	1	no choice	4–6 min.

In the exam

Make sure you read the front of the exam paper carefully.
Look at the exam cover opposite:
How is your exam paper different to this one?

Here are some hints on answering questions in the exam:

Answering 'structured' questions:

- Read the information at the start of each question carefully. Make sure you understand what the question is about, and what you are expected to do.

- Pace yourself with a watch so you don't run out of time. If you have spare time at the end, use it wisely.

- ***How much detail do you need to give?***
  The question gives you clues:
  - Give short answers to questions which start:
    '**State** . . .' or '**List** . . .' or '**Name** . . .'.
  - Give longer answers if you are asked to '**Explain** . . .' or '**Describe** . . .' or asked '**Why does** . . ?'.

- Don't explain something just because you know how to! You only earn marks for doing exactly what the question asks.

- Look for the marks awarded for each part of the question. It is usually given in brackets, e.g. [2]. This tells you how many points the examiner is looking for in your answer.

- The number of lines of space is also a guide to how much you are expected to write.

- Always show the steps in your working out of calculations. This way, you can gain marks for the way you tackle the problem, even if your final answer is wrong.

- Try to write something for every part of each question.

- Follow the instructions given in the question. If it asks for one answer, give only one answer. Sometimes you are given a list of alternatives to choose from. If you include more answers than asked for, any wrong answers will cancel out your right ones!

**▐▌ National
Examining Board**

SCIENCE:

BIOLOGY
Foundation Tier

4th June 9.30 a.m.

Time: 1 hour 30 minutes

Answer **ALL** the questions.

In calculations, show clearly how you work out your answer.

Calculators may be used.

Mark allocations are shown in the right-hand margin.

In what ways is your examination paper different from this?

Addiction When the body gets so used to a drug the person cannot do without it.

Aerobic respiration The release of energy from food with the help of oxygen.

Anaerobic respiration Respiration that occurs in the absence of oxygen.

Acid rain Rainfall containing dissolved gases like sulphur dioxide.

Alleles Different forms of the same gene.

Amino acids The small molecules that proteins are made of.

Antibiotics Drugs that kill bacteria.

Antibodies Proteins produced by white blood cells to destroy bacteria and viruses.

Antitoxins Chemicals made by white blood cells to destroy the toxins made by microbes.

Artificial selection The breeding of living organisms with characteristics useful to man.

Asexual reproduction Reproduction without sex cells.

Auxins A group of hormones that control plant growth.

Bile A chemical that neutralizes stomach acid and helps to break up fat molecules.

Biodegradable Materials that can be broken down by microbes.

Biological control Controlling pests using animals that normally prey upon them.

Biomass The mass of living material.

Cancer A disease caused when cell division goes out of control.

Chlorophyll A green substance that absorbs light energy.

Chloroplasts The structure in plant cells that contains chlorophyll.

Clones Genetically identical individuals.

CNS Central nervous system (brain and spinal cord).

Community All the living organisms living in one place.

Consumers Animals that eat other living things.

Decomposers Microbes that break down dead organisms.

Deforestation The cutting down of large areas of forest.

Depressant A drug that slows down the nervous system.

Detritivores Organisms (like earthworms) that shred up dead material.

Diabetes A disease in which the blood sugar level rises to a high level.

Digestion The breaking down of food into small soluble molecules.

Diffusion The movement of particles from an area of high concentration to an area of low concentration.

DNA Deoxyribonucleic acid, the chemical that chromosomes and genes are made of.

Ecosystem A group of living things and their environment.

Embryo transplantation Cloning animals by splitting embryos and putting them into host mothers.

Environmental variation Differences influenced by an organism's surroundings.

Enzymes Biological catalysts that speed up reactions in plants and animals.

Eutrophication Pollution of freshwater caused by an excess of nitrates.

Evolution The changes that occur in species over long periods of time.

Excretion Getting rid of waste products.

Fatty acids One of the two molecules that make up fats.

Fertilisers Chemicals added to crops to improve growth.

Fertility drugs Hormones that stimulate egg release.

Fossil fuels Fuels produced when animals and plants decayed millions of years ago.

Gametes Sex cells (e.g. sperms and eggs).

Genetic variation Differences caused by inheriting different genes.

Genetic engineering The transfer of human genes to bacterial cells.

Genes Small sections of chromosomes that control characteristics.

Gene therapy Replacing faulty genes.

Genetic disease A disease caused by faulty genes and passed on from one generation to the next.

Genotype The alleles that an organism has for any particular characteristic.

Glucose The simplest form of sugar.

Greenhouse effect The warming of the Earth as heat gets trapped in the atmosphere.

Greenhouse gas Gases (like carbon dioxide) that trap heat in the atmosphere.

Glucagon A hormone that helps to control blood sugar levels.

Glycerol The other molecule that makes up fat.

Glycogen The insoluble form in which glucose is stored in the body.

Guard cells Cells that open and close the stomata in leaves.

Habitat The place where an animal or plant lives.

Hormones Chemicals that help to coordinate the body's processes.

Human genome All the genes in a human cell.

Intensive farming Farming under very carefully controlled conditions.

Insulin A hormone that helps to control blood sugar levels.

Lactic acid A waste product of anaerobic respiration.

Limiting factor Something that slows down photosynthesis.

Menstrual flow The monthly loss of the womb lining from a woman's vagina.

Methane A 'greenhouse' gas released from cattle and rice fields.

Motor neurone   A cell that carries impulses from the CNS to muscles and glands.

Mutations   Changes that happen to genes.

Natural selection   How organisms best suited to their environment survive and breed.

Neurones   Nerve cells.

Nitrates   Chemicals containing nitrogen, often used in fertilisers.

Ovulation   The release of an egg from a woman's ovary.

Osmosis   The diffusion of water from a dilute to a more concentrated solution through a partially permeable membrane.

Oxygen debt   The oxygen needed to break down lactic acid.

Partially permeable membrane   A membrane that will let some molecules through but not others.

Pesticides   Chemicals that kill pests.

Phloem   The vessels that transport sugars in a plant.

Photosynthesis   How plants make food using carbon dioxide and water.

Pituitary gland   A gland at the base of the brain that produces hormones.

Population   A group of organisms of the same species.

Predators   Animals that catch and kill their own food.

Prey   Animals caught and killed by predators.

Producers   Green plants that make their own food.

Puberty   The time when boys and girls become sexually mature.

Receptors   Groups of cells that detect stimuli.

Reflex action   Rapid actions that you don't think about before doing them.

Response   A reaction to a stimulus.

Sensory neurone   A cell that carries impulses from receptors to the CNS.

Species   A population of living things that can breed and produce fertile offspring.

Stimuli   Changes in the environment.

Stomata   Tiny holes in the underside of leaves.

Sustainable development   Using the Earth's resources without destroying them.

Tissue culture   Cloning new plants from small tissue samples.

Transpiration   The loss of water vapour from leaves.

Urea   Made in the liver from the breakdown of excess amino acids.

Vaccine   Dead or weakened bacteria or viruses that cause antibodies to be made.

Withdrawal symptoms   The side effects that occur when a person gives up drugs.

Xylem   The vessels that carry water and minerals in a plant.

▶▶▶ Acknowledgements

I would like to thank my wife Jane and the rest of my family for all their support and encouragement during the writing of this book.

Special thanks are due to Gareth Williams for his advice, encouragement and for inspiring me to write *Science for you – Biology*.

I would also like to acknowledge the contribution made by Gareth Williams through his agreeing to the inclusion of certain illustrations from *Biology for You* in this book.

Particular thanks are due to Lawrie Ryan for his invaluable suggestions, advice and encouragement throughout the writing of the manuscript.

Thanks are also due to Michael Cotter, Beth Hutchins, Susannah Wills and Sarah Coulson for their comments and suggestions.

Finally I would like to dedicate this book to my late father Barrie whose memory inspires me still.

Acknowledgement is made to the following Awarding Bodies for their permission to reprint questions from their examination papers:

AQA	Assessment and Qualifications Alliance
AQA (NEAB)	Assessment and Qualifications Alliance
AQA (SEG)	Assessment and Qualifications Alliance
CGP	Coordination Group Publications
EDEXCEL	Edexcel Foundation
OCR	Oxford, Cambridge and RSA Examinations
WJEC	Welsh Joint Education Committee

Illustration acknowledgements

Ardea: 88b, 180; **BBC Picture Archives:** 184t; **Biophotos:** 71, 176b; **Bruce Coleman Collection:** 56b, 64, 86t, 108t, c, 111 all, 113, 114b, 115, 130, 136b, 155, 162b, 173, 186t, 187b, 198 both, 200 both; **Collections:** 6t, 68 both, 75b, 77, 187t; **Corbis UK Ltd:** 194t; **Ecoscene:** 61, 132b, 134, 177; **Eye Ubiquitous:** 34, 53, 80, 88t, 120, 137t, 162c; **Getty Images/Image Bank:** 143b; **Robert Harding Picture Library:** 15, 90t, 139t; **Hulton Getty:** 164; **Image State:** 133t, 136t, 156, 182b, 199b; **Impact Photos:** 16, 94t, 146t; **Jeff Moore** (jeff@jmal.co.uk): 58, 59b, 94b; **Natural Visions/ B Rogers:** 141t; **NHPA:** 86b, 176c, 186b, 191, 196; **Oxford Scientific Films:** 110b; **PA News Photos:** 32b; **Photos for Books:** 13, 17, 22, 26b, 60t, 91, 176t, 182t, 188; **Popperfoto:** 32t, 42t, 72 both, 110t, 136c, 181b, 190; **Science & Society Picture Library:** 7t, 140t; **Science Photo Library:** 6b, 7b, 9 both, 10 both, 11, 12, 20, 23, 26t, c, 30, 31, 38, 39, 42b, 46, 54 both, 56t, 59t, 60b, 63, 65, 69 both, 75t, 78, 82, 85, 90b, 92 both, 93 all, 96, 103, 104, 105, 106, 108b, 114t, 116 both, 123, 127, 132t, 133b, 137b, 139b, 140b, 141b, 142, 143t, 144 both, 145, 146b, 147, 157, 158, 161, 162t, 163 both, 166 both, 170, 171, 172, 174 both, 181t, 184b, 189, 192, 193, 194b, 199t, 201.

While every effort has been made to contact copyright holders, the publishers apologise for any omissions, which they will be pleased to rectify at the earliest opportunity.

Picture research by Liz Moore (lm@appleonline.net)